Pragmatic Circuits:
DC and Time Domain

Pragmatic Circuits: DC and Time Domain

William J. Eccles

ISBN (13 digits):978-3-031-79745-3   paperback

ISBN (13 digits):978-3-031-79746-0   ebook

DOI: 10.1007/978-3-031-79746-0

A Publication in the Springer series
SYNTHESIS LECTURES ON DIGITAL CIRCUITS AND SYSTEMS #2

Series Editor: Mitchell Thornton, Southern Methodist University

Series ISSN: 1932-3166   print
Series ISSN: 1932-3174   electronic

10 9 8 7 6 5 4 3 2 1

Pragmatic Circuits:
DC and Time Domain

William J. Eccles
Rose-Hulman Institute of Technology
Terre Haute, Indiana, USA

SYNTHESIS LECTURES ON DIGITAL CIRCUITS AND SYSTEMS #2

ABSTRACT

Pragmatic Circuits: DC and Time Domain deals primarily with circuits and how they function, beginning with a review of *Kirchhoff's* and *Ohm's Laws*, analysis of *d-c circuits* and *op-amps*, and the *sinusoidal steady state*. The author then looks at formal circuit analysis through nodal and mesh equations. Useful theorems like Thévenin's are added to the circuits toolbox. This first of three volumes ends with a chapter on design. The two follow-up volumes in the ***Pragmatic Circuits*** series include titles on ***Frequency Domain*** and ***Signals and Filters***.

These short lecture books will be of use to students at any level of electrical engineering and for practicing engineers, or scientists, in any field looking for a practical and applied introduction to circuits and signals. The author's "pragmatic" and applied style gives a unique and helpful "non-idealistic, practical, opinionated" introduction to circuits.

KEYWORDS:

circuit analysis, d-c circuits, time-domain circuits, sinusoidal steady state

Contents

CHAPTER 1

Introduction: What Do You Know?

Let's start all this with just a review of the basics before we tackle some of the more formal parts of electrical engineering. That's our goal in Chapter 1.

Do you remember the basic laws, good ol' Ohm and the two of Kirchhoff? Sure, no problem. How about dividers? Writing equations? Op-amps? Let's take a little time to review those. And while we are doing this, let's simplify things a bit by combining some steps into larger packages.

So off we go into what may be a review for you, a review that should get both of us marching in the same direction.

1.1 FUNDAMENTALS

Three very basic laws form the foundation of everything that we are going to do in this course. Here they are again, one at a time.

1.1.1 Ohm's Law

Stating Ohm's law is easy,

$$v = Ri.$$

It relates the voltage and current associated with a resistor, defined according to the *passive sign convention*. This convention, shown in Fig. 1.1, requires that the current enter the "+" end of the voltage.

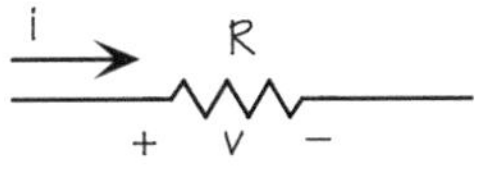

FIGURE 1.1: Passive sign convention.

We usually label resistances in ohms and use the standard SI symbol Ω to designate these. What happens when we turn things around and want i in terms of v is somewhat of a problem. The reciprocal of resistance is *conductance* and should be labeled with the unit of the *siemens*, whose standard SI symbol is S. Some folks persist in the pre-1962 unit of the *mho* and the upside-down omega for conductance. Conductances are often labeled with the letter G, as in, $G = 100$ mS.

The resistors in Fig. 1.2 all have the same relationship between voltage and current. All are labeled correctly according to modern SI convention.

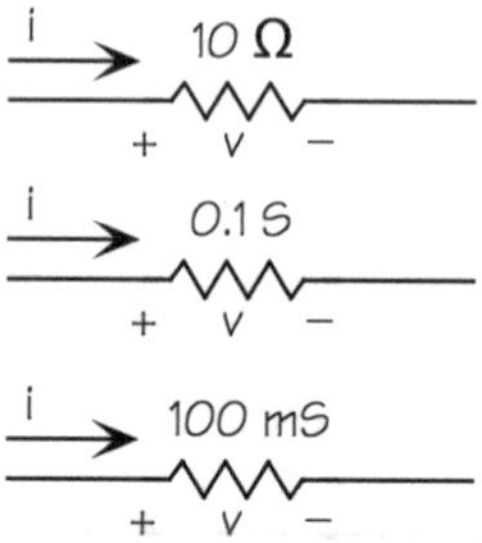

FIGURE 1.2:   Properly labeled resistors.

Ohm's law is extended to other components as well. Both the inductor and the capacitor introduce the element of time into our equations. But our circuit analysis is still based on "Ohm's law" defined for each of these elements.

The inductor (Fig. 1.3) stores magnetic energy. It has a voltage across it that is related to the time-rate-of-change of current:

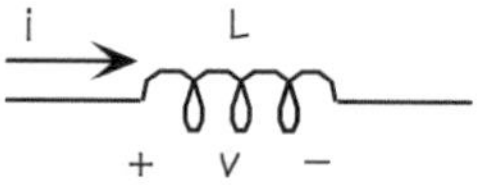

FIGURE 1.3:   Inductor.

$$v = L\frac{di}{dt}.$$

The capacitor (Fig. 1.4) stores electrical energy. It is the *dual* of the inductor, in that the effects of i and v in its defining equation are interchanged:

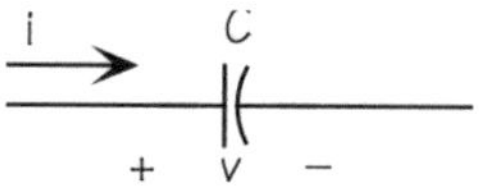

FIGURE 1.4:   Capacitor.

$$i = C\frac{dv}{dt}.$$

Ohm's law, whether for resistor, inductor, or capacitor, provides part of our equation set for a circuit, the part that we call the *element constraints*.

1.1.2   Kirchhoff's Current Law

Kirchhoff's current law states, in a very simple way, one of the most fundamental principles in electrical circuits:

> **The algebraic sum of the currents**
> **entering a node is zero**
> **at every instant of time.**

The word *node* in the law means more than just a place where elements join; we can read it as a *Gaussian surface*. If that bothers us, just think of that surface as any sensible closed surface. So Kirchhoff's current law (let's say KCL after this) applies to, say, the currents entering any enclosed volume.

But this enclosed volume thing brings up a new situation—the volume can be any size. How about the speed of light? After all, current shouldn't propagate through something faster than the speed of light. Here is where we bring into our discussion the concept of a *lumped* circuit element. For KCL to apply, we must think of our circuit elements as reduced to small lumps.

How small is a small lump? Small enough that the time for current to go through can be ignored. KCL doesn't treat relativity and we don't ask it to deal with circuits where this is a problem.

KCL provides the second set of constraints when we write circuit equations.

1.1.3   Kirchhoff's Voltage Law

Kirchhoff's voltage law (KVL for short) states another of our fundamental principles:

> **The algebraic sum of the voltages**
> **around any closed loop is zero**
> **at every instant of time.**

What has just been said about the relativistic effects applies here too, so we'll keep things small enough to stay out of that trouble. Our closed loop can be *any* closed loop; it doesn't have to follow the wires of our circuit.

KVL provides the third set of constraints when we write circuit equations.

1.1.4   Sources

Our sources provide either a voltage or a current to supply energy to our circuits. These sources are *ideal*, which means that what you see is what you get. If a source says it supplies 10 V, it does it for any current, even a gigaampere. We know that there aren't such things, but they are convenient for modeling real elements.

Fairly common standard ways of drawing these ideal sources are shown in Fig. 1.5.

The two sources on the left in the figure are voltage sources, providing the stated voltage between the two nodes to which they are connected. Nothing is said about the current through them, which can be any value in either direc-

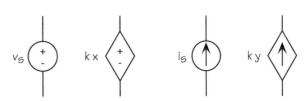

FIGURE 1.5:   Ideal sources.

tion. The leftmost voltage source provides v_s volts. The second one provides kx volts, where x is a voltage or current elsewhere in the circuit. In other words, the second one is *dependent* upon something else. So the first one is an *independent* voltage source, while the second is a *dependent* voltage source.

The two on the right are current sources, forcing the stated current to flow along the path they occupy. Nothing is said about the voltage across them, which can be any value in either direction. The one on the left provides i_s amperes. The one on the right provides ky amperes, where y is a voltage or current elsewhere in the circuit. Its current depends on something else. As with voltage source sources, the first is *independent* and the second is *dependent*.

1.1.5   Building Blocks

Every linear circuit that we construct can be modeled using these building blocks:

- Resistors, inductors, and capacitors;
- Ohm's law and the passive sign convention;
- Kirchhoff's current and voltage laws; and
- Ideal voltage and current sources, both independent and dependent.

In the next section we'll review some useful combinations of these elements to expand our toolbox and make us more efficient.

1.2   NICE TOOLS

Every circuits problem we can think of can be solved using only the basic tools that we have just seen: KVL, KCL, and Ohm's law for the various elements. Well…, almost. Which is a way of saying that we can end all this discussion right now. There is no need for the rest of the text…or the course, either.

If we had only the tools we consider basic, we'd find our work very time consuming. After all, you *can* take your car's engine apart with a large screwdriver and a set of box wrenches, but a few more specialized tools would certainly make the job easier.

Let's take a look at four of these "better" tools that you perhaps have already seen: series resistance, parallel resistance, voltage divider, and current divider.

1.2.1   Series Resistance

Resistors in series add. That's about all we need to say. So in the circuit in Fig. 1.6, the three resistors are equivalent to a single 60-Ω resistor.

How do we know for certain that resistors are in series? They must have exactly the same current flowing through each of them.

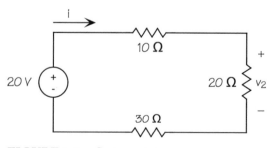

FIGURE 1.6:   Series resistors.

So the test for series is: Does the *entire* current that is flowing through one resistor flow through the second resistor?

1.2.2 Parallel Resistance

Resistors in parallel add reciprocally. To say it another way, conductances in parallel add. The three resistors in the circuit in Fig. 1.7 are in parallel, so we add them as conductances (in the equation below, I have chosen to convert *kilohms* to *millisiemens* to avoid lots of "ten-to-the," so 6 kΩ becomes 1/6 mS):

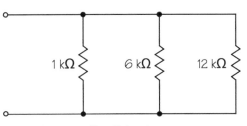

FIGURE 1.7:   Parallel resistors.

$$R_{total} = \frac{1}{\dfrac{1}{1}+\dfrac{1}{6}+\dfrac{1}{12}} = \frac{1}{\dfrac{15}{12}} = 0.8 \text{ k}\Omega = 800 \ \Omega.$$

Resistors are in parallel if they have exactly the same voltage across each of them. In other words, they must be connected to exactly the same pair of nodes. A simple test for parallel resistors is: Are *both* ends of the resistors connected to common nodes?

Conductances seem somewhat harder to work with than resistances, however, so we sometimes make use of the fact that the equation for *just two* resistors in parallel can be written in terms of resistance:

$$R_{parallel} = \frac{R_1 R_2}{R_1 + R_2}.$$

Two resistors in parallel combine as product over sum.

We can use this to redo the calculation of Fig. 1.7, combining one pair (say the 6- and the 12-kΩ resistors) and then combining the result with the 1-kΩ resistor (read the double bar as "in parallel with"):

$$R_{total} = 1\|(6\|12) = 1\left\|\left(\frac{6\times12}{6+12}\right)\right.$$

$$= 1\|4 = \left(\frac{1\times4}{1+4}\right) = 0.8 \text{ k}\Omega = 800 \ \Omega.$$

1.2.3   Voltage Divider

The voltage divider is a simple but very common circuit. The voltages across resistors in series are in the same ratio as the resistor values. So in the circuit in Fig. 1.8, we find v_o from the ratio of the 6-kΩ resistor to the sum of the two resistors:

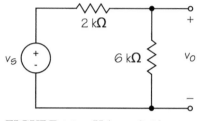

FIGURE 1.8:   Voltage divider.

$$v_o = v_s \frac{6}{6+2} = 0.75 v_s.$$

A common electrical device is the *potentiometer*, although anybody who calls it that is suspect. The variable resistor is usually called a *pot*. A fixed resistance is fitted with a slider that contacts the resistance, thereby dividing the resistance into two parts.

In Fig. 1.9, the fixed resistance is 10 kΩ. The slider, depicted by an arrow, is a fraction a from the lower end of the resistance. The voltage out is therefore the fraction a ($0 \le a \le 1$) of the input,

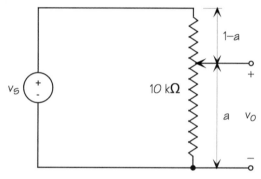

FIGURE 1.9:   Pot(entiometer).

$$v_o = a v_s.$$

1.2.4   Current Divider

The current divider (Fig. 1.10) consists of resistors in parallel. The currents through them divide in the ratio of their *conductances*. But the most useful form of the current divider is with just two resistors in parallel. Here, the currents divide much like voltages divide in the voltage divider, but there's one difference—the numerator is the *other* resistor:

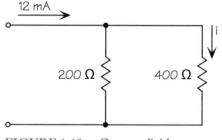

FIGURE 1.10:   Current divider.

$$i_{400} = 12 \frac{200}{200 + 400} = 4 \, \text{mA}.$$

Note that the current through the 400-Ω resistor requires the 200-Ω value in the numerator. And don't forget that this works only for *two* resistors in a current divider.

1.3 SOLVING—ALL THE WAY

Even though we have neat tools such as the voltage divider, we don't need any of them to be able to write a complete mathematical description of a circuit and "solve it." Kirchhoff's laws and Ohm's law are sufficient.

My purpose in this section is to "solve" several circuits using this basic approach. I call it "firing the big gun" because we write every equation we can. At the same time, we'll make a run through the DC circuits, the time domain, and the sinusoidal steady state. This will probably be a review of some of what you already know.

1.3.1 DC Circuit

Let's start with the DC circuit of Fig. 1.11. Our goal is to find the output voltage v_o.

The first step in the "big gun" solution technique is to label all of the voltages and all of the currents *everywhere* in the circuit. I've done that in Fig. 1.12. Note that I have assigned a variable to the voltage across a voltage source and also to the current through the current source.

Since every element has two labeled variables, one current and one voltage, there are twice as many variables as there are elements. So for our six elements we have 12 variables. That means 12 equations in 12 unknowns.

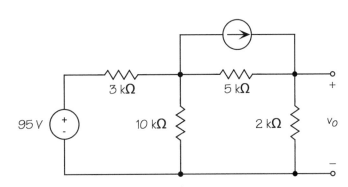

FIGURE 1.11: DC circuit to solve.

We must use all of the basics. Let's write the Ohm's-law constraints first, then the KCL constraints, and finally the KVL constraints.

There are six elements, so there are six Ohm's-law constraints (including the source constraints):

$$v_{s1} = 95, v_3 = 3 \times i_3, v_{10} = 10 \times i_{10}, v_5 = 5 \times i_5, i_{s2} = 4, v_o = 2 \times i_2.$$

The units are volts, kilohms, and milliamperes, which are self-consistent.

Now for KCL. The circuit has four nodes. We can write KCL constraints at any three; writing one at the fourth would yield a redundant equation. I'll choose the three across the top:

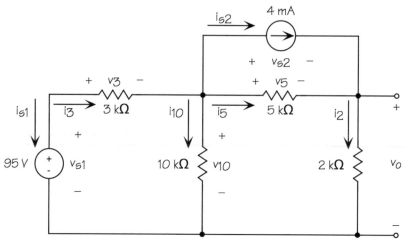

FIGURE 1.12:   Completely labeled DC circuit.

$$-i_{s1} - i_3 = 0,$$
$$+i_3 - i_{10} - i_5 - i_{s2} = 0,$$
$$i_5 + i_{s2} - i_2 = 0.$$

Finally, KVL. There are three meshes ("smallest loops"), so there are three equations:

$$-v_{s1} + v_3 + v_{10} = 0,$$
$$-v_{10} + v_5 + v_o = 0,$$
$$-v_5 + v_{s2} = 0.$$

Whee! That's 12 equations, which is what I needed. Solving these yields

$$v_o = 20 \text{ volts.}$$

1.3.2   Circuit with Dependent Source

Dependent sources seem to add complexity to a circuit. But when we are firing the "big gun," they merely add new constraints. Fig. 1.13 contains a dependent source. I've put labels on the variables.

This circuit has four elements. Since each should have two variables associated with it, there should be eight equations with eight unknowns. This time, though, I have cheated a bit. In three instances I have "used my head" and eliminated a variable:

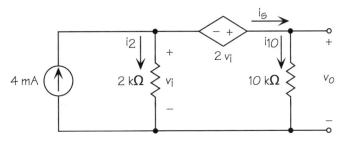

FIGURE 1.13: Circuit with dependent source.

- The current upward through the 4-mA source is 4 mA, so I didn't give that a separate variable name.
- The voltage across the current source is v_i, so its voltage doesn't need a separate label.
- The dependent source's voltage is already given in terms of v_i, so I avoided another variable.

The result will be five equations in five unknowns: two element constraints (Ohm's law) for the resistors, two KCL equations at the two upper nodes, and one KVL equation around the mesh on the right:

$$v_i = 2i_2,$$
$$v_o = 10i_{10},$$
$$4 - i_2 - i_s = 0,$$
$$i_s - i_{10} = 0,$$
$$-v_i - 2v_i + v_o = 0.$$

Note here that I have used volts, kilohms, and milliamperes. These are self-consistent and save writing several ten-to-the terms. The solution of this set yields

$$v_o = 15 \text{ volts.}$$

1.3.3 Time-Domain Circuit

Firing the "big gun" at a circuit in the time domain means only that the circuit elements now involve the independent variable time. Fig. 1.14 includes an inductor and a capacitor.

I have labeled enough voltages and currents to be able to write element constraints for all five passive elements:

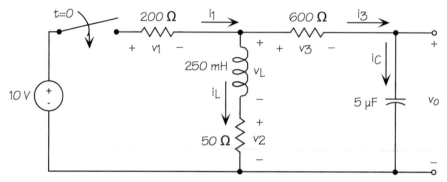

FIGURE 1.14:   Time-domain circuit.

$$v_1(t) = 200i_1(t),$$

$$v_L(t) = 0.25\frac{di_L(t)}{dt},$$

$$v_2(t) = 50i_L(t),$$

$$v_3(t) = 600i_3(t),$$

$$i_C(t) = 5 \times 10^{-6}\frac{dv_o(t)}{dt}.$$

I will also write two KCL equations at the two top nodes and two KVL equations around the two meshes:

$$i_1(t) - i_L(t) - i_3(t) = 0,$$
$$i_3(t) - i_C(t) = 0;$$
$$-10 + v_1(t) + v_L(t) + v_2(t) = 0,$$
$$-v_2(t) - v_L(t) + v_3(t) + v_o(t) = 0.$$

Two more equations will finish this—the initial conditions for the energy-storage elements:

$$i_L(0) = 0,$$
$$v_o(0) = 0.$$

These are zero because the switch has been open for a long time before $t = 0$ and hence there is no initial energy storage. Any energy in the form of inductor current (inductor energy = $\frac{1}{2}Li^2$) or capacitor voltage (capacitor energy = $\frac{1}{2}Cv^2$) that could have been there a long time ago has dissipated.

The result, after some effort on the part of Maple, is

$$v_o(t) = 2.00 + 3.529e^{-364.9t} - 5.529e^{-685.1t} \text{ V}, \text{ } t \geq 0,$$

which is plotted in Fig. 1.15.

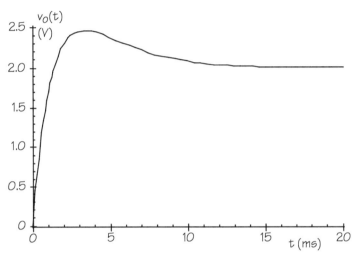

FIGURE 1.15:   Time-domain circuit output.

1.3.4   Sinusoidal Steady State

The "sinusoidal steady state" sounds like something that only psychiatrists could treat. But you probably aren't a psychiatrist. You still know that this version of a circuit considers only the nontransient effects of sine-wave sources.

When you first saw sinusoidal steady-state analysis, you probably worked with circuits with sine waves at fixed frequencies. You defined *impedance* as the sinusoidal steady-state equivalent of resistance. Impedance and its reciprocal, *admittance*, allow us to consider resistors, capacitors, and inductors under one umbrella.

Let's continue with the circuit we've just used, but now apply a source that looks like the local power line: 120 V rms at 60 Hz. The relabeled circuit is shown in Fig. 1.16.

Note that I have labeled the currents and voltages with capital letters. Some authors distinguish between time-domain variables and sinusoidal steady-state variables via this convention. (I'll try to consistently do this.)

We start by defining the radian frequency ω:

$$\omega = 2\pi 60 = 377 \text{ radians/second.}$$

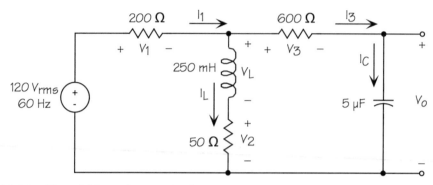

FIGURE 1.16:   Sinusoidal steady-state circuit.

Now define the impedance of the middle (vertical) branch and of the capacitor. Remember that the impedance of an inductor is $j\omega L$ and of the capacitor, $1/j\omega C$:

$$Z_m = j0.25\omega + 50 = 50 + j94.248 \ \Omega,$$
$$Z_c = 1/(j\omega 5 \times 10^{-6}) = -j530.52 \ \Omega.$$

Once we've done this, the "big gun" approach produces element constraints as before, except that there are now only four passive elements (Z_m includes the inductor and the resistor in series):

$$V_1 = 200I_1,$$
$$V_m = Z_m I_L,$$
$$V_3 = 600I_3,$$
$$V_o = Z_c I_c.$$

The KCL and KVL constraints are the same as before, except for the capital letters and the combination of the voltages V_L and V_2 into V_m:

$$I_1 - I_L - I_3 = 0,$$
$$I_3 - I_C = 0;$$
$$-10 + V_1 + V_m = 0,$$
$$-V_m + V_3 + V_o = 0.$$

The result of solving these equations is the *phasor*

$$V_o = 30.441 - j6.861 \text{ volts,}$$

which can be written in polar form as

$$V_o = 31.205 \angle -12.70° \text{ volts.}$$

These two results are the sinusoidal steady-state response of this circuit. They are written in the complex-number notation that we use when we aren't concerned with the time-domain response, a form that we call a *phasor*. The frequency is only implied here; we have to go to the drawing to see that it's 60 Hz.

Converting this result to the time domain requires us to reintroduce the frequency. So we are converting from the *phasor domain* to the *time domain*:

$$v_o(t) = 31.205 \cos(377t - 12.70°) \text{ volts.}$$

(I arbitrarily chose the cosine as the reference. There's no specific standard for this, but *cosine* is more common in circuits.)

Remember that it is totally incorrect to equate in any way V_o and $v_o(t)$; the first is the frequency-domain result (a phasor), the second is the time-domain result.

1.3.5 Power and Energy
Power is the product of the voltage across an element and the current through that element:

$$p = vi.$$

Its unit is the *watt*, abbreviated W. If we define voltage and current according to the passive sign convention (Fig. 1.1), then a positive number for power means that the element is absorbing power.

For a resistor,

$$v = Ri,$$
$$p = vi = Ri^2 = v^2/R.$$

If v and i are time varying, then so is power.

Energy is the accumulation of power over a period of time. For example, the energy delivered to an element between t_1 and t_2 is

$$w = \int_{t_1}^{t_2} p(t)dt.$$

Its unit is the *joule*, abbreviated J.

Suppose that a 100-Ω resistor has 120 V applied to it. What is the energy delivered to this resistor in 25 s?

$$p = v^2/R = 120^2/100 = 144 \text{ W},$$
$$w = \int_0^{25} 144 dt = 3600 \text{ J}.$$

If we do all this for the inductor, we get

$$p = vi = L\frac{di}{dt}i.$$

A time varying current means that the power delivered to the inductor is time varying. If the current is not time varying, then the power is zero because di/dt is zero.

An inductor stores energy, which we compute from the power delivered to the inductor:

$$w_L(t) - W_L(0) = \int_0^t L\frac{di}{dt}i dt = \int_{i(0)}^{i(t)} Li \, di = \frac{1}{2}L\left(i(t)^2 - i(0)^2\right).$$

This says that the energy stored in an inductor is determined by the current through the inductor at any time. We usually pick zero energy to occur at zero current. If we start at $t = 0$ with zero current, then the total energy stored in the inductor at time t is

$$w_L(t) = \frac{1}{2}Li^2.$$

Now do all this for the capacitor:

$$w_C(t) = \frac{1}{2}Cv^2.$$

In circuits involving sinusoids, power and energy are time varying. Generally that isn't what we want to know about a circuit, so we define *average power* based on the *root-mean-square* (rms) values of voltage and current:

$$v_{rms} = \frac{1}{period} \int_{period} v^2(t) dt,$$
$$P_{ave} = V_{rms} I_{rms}.$$

For a *sine wave*, v_{rms} works out to be $1/\sqrt{2}$ or 0.707 of the peak value (amplitude) of the sine wave.

One reason for using rms values is that a resistor will absorb the same power when we apply either an rms voltage or a DC voltage of the same numerical value. For example, if we apply a 120-V_{rms} sine wave to a 100-Ω resistor, the power absorbed by the resistor is 144 W, the same as in the DC case earlier in this section.

1.4   OP-AMPS

The operational amplifier is usually called an *op-amp* because its real name is too long to say. Fig. 1.17 shows an op-amp with all its terminals labeled.

The device requires power, shown in the figure as V_{cc}. But when we draw most circuits, we ignore the power terminals, since they rarely enter our analysis of the circuit around the op-amp.

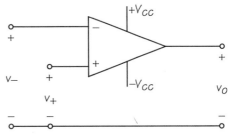

FIGURE 1.17:   Operational amplifier.

The two inputs are labeled v_- and v_+. The gain of the amplifier is very high, typically of the order of 10^5 or 10^6. If this gain is called A, then the output of the op-amp is the gain times the difference of the inputs:

$$v_o = A(v_+ - v_-).$$

(*A* is called the *open-loop* gain, which is the gain straight through the op-amp without any external circuitry attached to it.)

The output v_o will not go beyond $\pm V_{cc}$ because the amplifier *saturates*. Actually, the output can't quite reach that level but instead falls a little short. In typical semiconductor op-amps like the LM 741 and LM 747, this shortage is about 1.5 V.

1.4.1   Ideal Op-amp

The current that flows into the inputs of the modern op-amp is very, very tiny, often micro- or nanoamperes. In most of our circuits we don't have currents that small (milliamperes is "small" for us). Therefore we can, as a decent approximation, ignore the op-amp's input currents when analyzing circuits.

Moreover, the open-loop gain A is so large that the input voltage must be minuscule compared to the output voltage. In most of our circuits, the output voltage is in the same range as the rest of the voltages in the circuit. Therefore the input voltage (the difference between v_+

and v_-) is essentially zero. So a decent approximation is to consider this voltage difference to be zero.

We call this negligible voltage difference a *virtual short*. The voltage between these two terminals is assumed to be zero, as it would be if the terminals were shorted together? No. current can flow over this "short circuit", though.

Finally, the op-amp involves a power supply that is not shown as a part of our circuit. Because of this, we cannot calculate the current entering the output of the op-amp. To say this another way, we can't write KCL at the output node of the op-amp until we have finished the rest of our analysis of the circuit.

Fig. 1.18 shows the ideal op-amp with "decent approximations" we have just described.

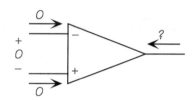

FIGURE 1.18: Ideal op-amp.

1.4.2 Circuit Example

Let's analyze one circuit, the inverter shown in Fig. 1.19. The op-amp is ideal, or at least we are going to assume that it is, and the power-supply inputs are not shown.

The analysis proceeds as follows:

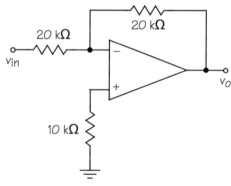

FIGURE 1.19: Op-amp inverter.

- Consider first the vertical 10-kΩ resistor. Since our assumption is that no current will flow into the + terminal, the current through the 10-kΩ resistor is 0. Therefore the voltage at the + terminal is the ground voltage, or 0.
- Use the assumption that the voltage difference between the + and − terminals is 0. Therefore the voltage at the − terminal is the same as at the + terminal, namely, 0.
- The voltage across the 20-kΩ input resistor is $v_{in}-$ 0; the voltage across the 20-kΩ feedback resistor is v_o- 0.
- The current flowing into the − input terminal is 0 because of the ideal assumptions.
- Applying KCL at the − input terminal requires the sum of three currents:

$$+\frac{v_{in}-0}{20}+\frac{v_o-0}{20}-0=0,$$

which solves to yield

$$v_o = -\frac{20}{20}v_{in} = -v_{in}.$$

The result of our analysis is that the output is merely a negative copy of the input, provided we don't try to push the output too close to the power-supply voltages.

Do you wonder what the 10-kΩ resistor is doing there? For the ideal op-amp, the answer is: Nothing! So why waste a part? Well, it provides *bias compensation*. Gee, really?

When we use an op-amp, the input currents aren't exactly zero, although they are tiny enough to neglect in the external circuit. But there are currents into each input. If these are not essentially equal, they produce a voltage difference that is also amplified by the op-amp. So these *bias currents* produce a DC offset in the output voltage, an offset that may not be acceptable in our application.

The DC offset produced by unbalanced bias currents can be controlled by making sure that each of the inputs "sees" the same resistance. In our example, the − input is "looking" into two 20-kΩ resistors that, to the − input at least, appear to be in parallel. The value of two 20-kΩ resistors in parallel is 10 kΩ. If we include 10 kΩ in the path to the + input, we reduce considerably the difference between the input currents and hence their effect on the output voltage.

1.5   ADDITIONAL EXAMPLES

Each chapter has a set of examples that review most of the major points developed in the chapter. Each one will be worked through to completion, although at times intermediate steps such as the Maple solution will be omitted.

1.5.1   Example I

Find the power delivered to the 200-Ω resistor in the circuit in Fig. 1.20.

Rather than firing the big gun at this problem, I'll use some of the simplifying tools from our tool box. I'm going to do parallel resistance and then series resistance. Next I'll find the source current and use the current divider to get the current in the 200-Ω resistor. From that, I'll get power.

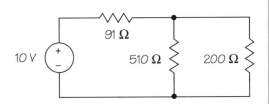

FIGURE 1.20:   Circuit for Example I.

First, let's combine in parallel the two resistors on the right:

$$\frac{1}{1/510 + 1/200} = 143.7 \ \Omega.$$

Now combine this result and the input resistor in series:

$$91 + 143.7 = 234.7 \, \Omega.$$

The source current is the source voltage divided by this total resistance:

$$10/234.7 = 42.61 \, \text{mA}.$$

The current divider says that the fraction of the current through one of a pair of parallel resistors is the *other* resistor divided by the sum of the resistors:

$$I_{200} = 42.61 \times \frac{510}{510 + 200} = 30.61 \, \text{mA}.$$

Finally, use *I*-squared-*R* for the power:

$$P_{200} = 200 \times I_{200}^2 = 187.4 \, \text{mW}.$$

The result is 187.4 mW.

Is this method of solution an improvement over firing the big gun? Maybe! It really depends on how you see the problem. This method does not require solving a bunch of equations, and all the individual steps are simple "calculator" ones.

1.5.2 Example II

The circuit of Fig. 1.21 has been sitting with the switch open for a long time before $t = 0$. Find v_L, the voltage across the inductor, for $t > 0$. The source voltage $v_s(t) = 20$ V.

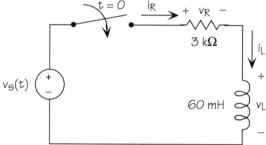

FIGURE 1.21: Labeled circuit for Example II.

The circuit as shown is labeled for firing the big gun, which is what I'm going to do. There are two elements (excluding the source), so I need four equations. Two of them are element equations, one is Kirchhoff's current law applied at the node at the upper right, and one is Kirchhoff's voltage law applied around the whole loop.

Since there is an energy-storage element involved, I need an initial condition. The inductor's stored energy is proportional to the square of the current, so this initial value needs to be of the initial current. Since the circuit has been dead for a long time before $t = 0$, this initial current is 0.

$$v_R = 3 \times 10^3 i_R,$$

$$v_L = 60 \times 10^{-3} \frac{di_L}{dt},$$

$$v_s = 20,$$

$$-i_R + i_L = 0,$$

$$-v_s + v_R + v_L = 0,$$

$$i_L(0) = 0.$$

I solved the equations in Maple to get

$$v_L = 20e^{-50000t} \text{ V for } t > 0.$$

Note that the voltage jumps up to 20 at $t = 0$ and then decays to 0 with a time constant of $1/50,000 = 20$ μs. Does this make sense? A long time after $t = 0$, currents and voltages are steady again. If the current is steady, its derivative is 0, so the voltage decays to 0. (The steady current is $20/3 = 6.67$ mA.)

1.5.3 Example III

Let's convert the previous circuit (without the switch) to the phasor domain. $v_s(t) = 20 \cos(40,000t)$ V. The modified circuit is shown in Fig. 1.22.

I'll convert the inductor to an imped-ance and then combine this impedance and the resistor into the total impedance of the loop:

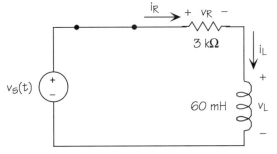

FIGURE 1.22: Labeled circuit for Example III.

$$Z_L = j \times 40,000 \times 60 \times 10^{-3} = j2400 \, \Omega,$$

$$Z_{total} = 3000 + j2400 \, \Omega.$$

The current in the loop (clockwise) is

$$I_{loop} = \frac{20\angle 0°}{Z_{total}} = 4.065 - j3.252 \text{ mA.}$$

Now I find the voltage across the inductor using Ohm's law:

$$V_L = Z_L \times I_{loop} = 7.805 + j9.756 = 12.49\angle 51.3° \text{ V}.$$

This returns to the time domain as

$$v_L(t) = 12.49\cos(40{,}000t + 51.3°) \text{ V}.$$

Note that the phase angle is stated in degrees. Most electrical engineers use degrees rather than radians, even though this is "apples + oranges."

1.5.4 Example IV

The switch in the circuit (Fig. 1.23) has been closed for a long time before $t = 0$, connecting a 15-volt source to the circuit. The switch opens at $t = 0$, leaving the circuit to "run on its own." Find $v_o(t)$ for $t > 0$.

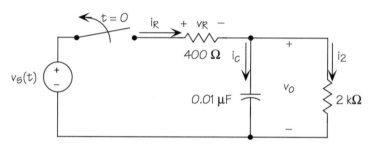

FIGURE 1.23:   Labeled circuit for Example IV.

First, though, we need the initial value of the voltage across the capacitor, becausse that voltage represents the capacitor's stored energy $(\frac{1}{2}Cv^2)$. Because everything is steady before $t = 0$, the voltage v_o is steady, so its derivative is 0. Hence the capacitor current is 0. That leaves just a voltage divider to determine the initial value of v_o:

$$v_o(0) = 15\frac{2000}{2000 + 400} = 12.5 \text{ V}.$$

Now that we have that, just fire the big gun, which deals with just two elements (the 400-Ω input resistor has no current through it after $t = 0$ because the switch is now open):

$$i_C = 0.01 \times 10^{-6}\frac{dv_o}{dt},$$
$$v_o = 2000i_2,$$
$$i_C + i_2 = 0,$$
$$v_o(0) = 12.5.$$

Solving these, we get

$$v_o(t) = 12.5e^{-50,000t} \text{ V for t} > 0.$$

As we should expect, the voltage dies to 0 because the stored energy runs out. How fast does this "run out?" The time constant is 1/50,000 = 20 μs. After five time constants, less than 1% of the stored energy remains. So we could say that the circuit is essentially dead in 100 μs.

1.5.5 Example V

In the circuit of Fig. 1.24, find the sinu-soidal steady-state voltage $v_o(t)$ for $v_s(t)$ = 15 cos(100,000t + 45°) V.

The voltage source in the phasor domain is

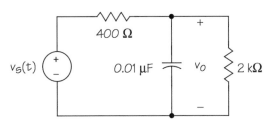

FIGURE 1.24: Labeled circuit for Example V.

$$V_s = 15\angle45° = 15e^{j45\pi/180} \text{ V.}$$

The capacitor becomes

$$Z_C = 1/(j100 \times 10^3 \times 0.01 \times 10^{-6}) = -j1000 \, \Omega.$$

Now find the impedance of the capacitor and the 2-kΩ resistor in parallel, then compute the voltage using a voltage divider:

$$Z_o = \frac{1}{1/2000 + 1/-j1000} = 400 - j800 \, \Omega,$$

$$V_o = V_s \frac{Z_o}{400 + Z_o} = 10.61 + j5.30 = 11.86\angle26.6° \text{ V.}$$

The sinusoidal steady-state result in the time domain is

$$v_o(t) = 11.86\cos(100,000t + 45° + 26.6°)$$
$$= 11.86\cos(100,000t + 71.6°) \text{ V.}$$

1.5.6   Example VI

Fig. 1.25 is a collection of parts that perhaps has too many elements. Our goal is to reduce that number by combinations of elements in series or in parallel.

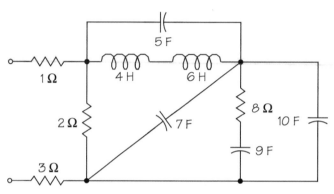

FIGURE 1.25:   Example VI before combining elements.

There are just three combinations possible:

- The two inductors are in series. Since inductors add like resistors add in series, their combination yields 4 + 6 = 10 H.
- Two capacitors are in parallel, the 7-F and 10-F capacitors. Neither of the other two capacitors combines; to be in parallel, the elements must share a common node at both ends. Capacitors in parallel add like resistors in series, so the combination yields 7 + 10 = 17 F.
- Two resistors are also in series, although they are separated in the circuit. The 1-Ω and 3-Ω resistors have identical current through them. In series, they become 1 + 3 = 4 Ω. The 2-Ω resistor is not in series with these because it does not receive the same identical current.

The resultant circuit is shown in Fig. 1.26.

1.5.7   Example VII

Use the big gun to find v_o in the circuit of Fig. 1.27.

In writing the equations, I have used labels that match the values of the elements. All voltages are positive at the left or top; all currents observe the passive sign convention. The big gun equations are

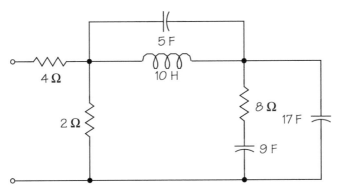

FIGURE 1.26: Example VI after combining elements.

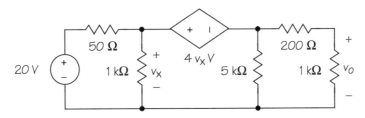

FIGURE 1.27: Example VII.

$$v_{50} = 50i_{50}, \qquad -i_{50} + i_x + i_{dep} = 0,$$
$$v_x = 1000i_x, \qquad -i_{dep} + i_5 + i_{200} = 0,$$
$$v_{dep} = 4v_x, \qquad -i_{200} + i_0 = 0,$$
$$v_5 = 5000i_5, \qquad -20 + v_{50} + v_x = 0,$$
$$v_{200} = 200i_{200}, \qquad -v_x + v_{dep} + v_5 = 0,$$
$$v_o = 1000i_o, \qquad -v_5 + v_{200} + v_o = 0.$$

Solving these yields

$$v_o = -55.87 \text{ V}.$$

Note that the output is much larger in magnitude than the source. This is the effect of the dependent source, something that you'll see when you model such circuits as transistor amplifiers.

1.5.8 Example VIII

Using variables labeled as I did in the previous example, I want $v_o(t)$ for $t > 0$ in the circuit of Fig. 1.28. Note that the source is a step, which has been dead forever before $t = 0$, so initial conditions are all 0.

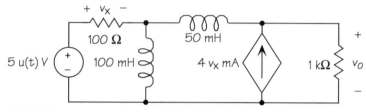

FIGURE 1.28: Example VIII.

The big gun equations in differential form are

$$v_x = 100 i_x,$$

$$v_{L1} = 100 \times 10^{-3} \frac{di_{L1}}{dt},$$

$$v_{L5} = 50 \times 10^{-3} \frac{di_{L5}}{dt},$$

$$v_o = 1000 i_o,$$

$$-i_x + i_{L1} + i_{L5} = 0,$$

$$-i_{L5} - 4 \times 10^{-3} v_x + i_o = 0,$$

$$-5 + v_x + v_{L1} = 0,$$

$$-v_{L1} + v_{L5} + v_o = 0,$$

$$i_{L1}(0) = 0, \ i_{L5}(0) = 0.$$

The solution of this is

$$v_o(t) = 5.391 e^{-659.2t} - 5.391 e^{-30341t} \text{ V for } t > 0.$$

1.5.9 Example IX

For the circuit of Fig. 1.29, apply the big gun in the phasor domain to find $v_o(t)$ in the sinusoidal steady state.

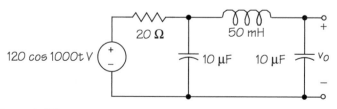

FIGURE 1.29: Example IX.

First, I need the impedances of the inductor and the capacitors:

$$\omega = 1000,$$
$$Z_C = 1/(j \times 1000 \times 10 \times 10^{-6}) = -j100 \, \Omega,$$
$$Z_L = j \times 1000 \times 50 \times 10^{-3} = j50 \, \Omega.$$

Now I'll write the big gun equations using labeling as I've done before:

$$V_{20} = 20 I_{20},$$
$$V_C = Z_C I_C,$$
$$V_L = Z_L I_L,$$
$$V_o = Z_C I_o,$$
$$-120 + V_{20} + V_C = 0,$$
$$-V_C + V_L + V_o = 0,$$
$$-I_{20} + I_C + I_L = 0,$$
$$-I_L + I_o = 0.$$

The solution in the phasor domain is

$$V_o = 176.5 - j105.9 = 205.8 \angle -31.0° \, \text{V}.$$

The sinusoidal steady-state time-domain result is

$$v_o(t) = 205.8 \cos(1000t - 31.0°) \, \text{V}.$$

1.5.10 Example X

Now let's do some op-amp circuits. Fig. 1.30 is an inverter. I'll label the variables as I have in previous examples and then write the big gun equations.

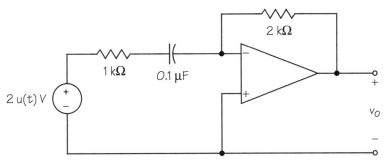

FIGURE 1.30:   Example X.

The initial value of the capacitor voltage is 0 because the source has been dead for a long time before $t = 0$.

In writing the equations, I'll make use of the ideal properties of the op-amp: zero current flowing into each input terminal and zero voltage difference between those terminalsl:

$$v_1 = 1000i_{in},$$

$$i_{in} = 0.1x10^{-6}\frac{v_c}{dt},$$

$$v_2 = 2000i_2,$$

$$-2 + v_1 + v_C - 0 = 0,$$

$$v_2 + v_o = 0,$$

$$-i_{in} + 0 + i_2 = 0,$$

$$v_C(0) = 0.$$

The solution for $v_o(t)$ is an exponential starting at −4 V that dies out with a time constant of $1/10,000 = 100$ μs:

$$v_o(t) = -4e^{-10000t} \text{ V for } t > 0.$$

1.5.11 Example XI

The next circuit (Fig. 1.31) is a noninverter to be analyzed in the phasor domain. Before starting, though, I note that the ideal op-amp has zero current flowing into its terminals, as well as a zero voltage difference between those terminals. Hence there is no current through the 200-Ω resistor; I can ignore it.

First, convert the circuit to the phasor domain, which here means finding the impedance of the capacitor. Then, using labeling as I've done before, write the big gun equations:

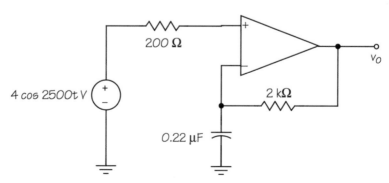

FIGURE 1.31:   Example XI.

$$\omega = 2500,$$
$$Z_C = 1/(j \times 2500 \times 0.22 \times 10^{-6}) = -j1818 \, \Omega.$$

$$V_{22} = Z_C I_{22},$$
$$V_2 = 2000 I_2,$$
$$-4 + V_{22} = 0,$$
$$-V_{22} + V_2 + V_o = 0,$$
$$I_{22} + I_2 = 0.$$

The phasor-domain solution is

$$V_o = 4.0 + j4.4 = 5.946 \angle 47.7° \, \text{V},$$

which gives a steady-state time-domain result of

$$v_o(t) = 5.946 \cos(2500t + 47.7°) \, \text{V}.$$

1.5.12 Example XII

Let's do the previous example once more, but this time let's use previous knowledge. We know that this circuit is a noninverter. We know that the gain of the noninverter is $1 + R_f/R_1$, which applies to impedances as well.

In the previous example, I found the impedance of the capacitor to be $-j1818 \, \Omega$. So the gain of the circuit is

$$1 + \frac{2 \times 10^3}{-j1818}.$$

Multiply this by the magnitude of the source voltage and the result is the same as before.

1.6   CIRCUIT DESIGN EXAMPLE

Now that we know what we know, let's design a bank thermometer! Huh? Here? Now? Well, yes and no. We'll design the circuit that gets a voltage proportional to temperature and sends it to an analog-to-digital converter. Then we'll let the digital folks take over to create the digits for the display and send those to the signmaker.

Here are our specifications:

- The input is to be a signal representing the outdoor temperature from an appropriate sensor.
- The output is to be a voltage that will be sent to an analog-to-digital (A/D) converter that the digital designer has already chosen.
- The digital coding from the A/D converter is a byte, where 0000 0000 represents − 128°F and 1111 1111 represents +127°F.
- The A/D converter is linear. An input of 0 V produces the code 0000 0000; +5 V produces 1111 1111. Its input impedance is about 5 kΩ.

I have chosen to divide the problem into several design steps, each influencing the next:

1. choice of temperature sensor and development of sensor circuit,
2. translation of sensor output to the required A/D converter input,
3. specification of power supplies, and
4. integration of these parts into a complete system.

Off we go!

1.6.1   Sensor

A search of catalogs and manufacturers' literature will uncover a number of devices whose output is nicely proportional to temperature. One group of these includes LM 134, 234, and 334. While LM 134 and LM 234 are specified as "true temperature sensors," I happen to have LM 334 available.

LM 334 is guaranteed only over the range of temperature from 0°C to +70°C, but it should be good enough for a typical bank thermometer sign. If my circuit design works out well, I might later change the design to use the somewhat wider temperature range available with one of the other devices.

LM 334 is actually a current source that operates on a voltage input range of +2 to +25 V. It can produce up to 10 mA. The basic circuit for LM 334 is shown in Fig. 1.32, along with the pinout diagram.

To connect the device into a circuit, we connect pin 2 to a stable power supply of from +2 to +25 V. Pin 3 is connected to the load we wish to supply current for. The resistor for R_{set} is connected between pins 1 and 3.

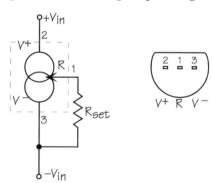

FIGURE 1.32:   LM 334 two-terminal current source.

The data sheet says the relationship between the current supplied (called i_{Tset}) and R_{set} is given by the relationship

$$R_{set} = \frac{0.0677}{i_{Tset}} \, \Omega \text{ at } 25°C.$$

Once this resistance value is chosen, i_T is linear with temperature in K, so $i_T = 0$ at 0 K.

Now comes the choice of R_{set}. Let's start by noting the temperature range in K. The lowest design value is $-128°F = 184.11 \, °K$; the highest is $+127°F = 325.78$ K. We also need $25°C = 298$ K.

The maximum current for LM 334 is 10 mA. I will choose to set the current for the highest temperature to 9 mA so we don't go to the maximum. A graph makes this clear in Fig. 1.33.

Because the graph is linear, we get i_{Tset} and R_{set} directly:

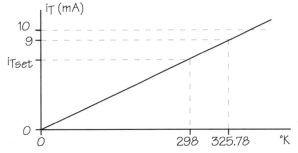

FIGURE 1.33:   Graph for I_{Tset} at 9 mA maximum.

$$i_{Tset} = 9\frac{298}{325.78} = 8.2325 \text{ mA},$$

$$R_{set} = \frac{0.0677}{8.2325} = 8.2235 \, \Omega.$$

Hmmmmmmmmmm, that's a rather low resistance value, although it is within the acceptable range for LM 334. I think I'll reduce the current greatly by choosing i_{Tset} at 25°C (298 K) to be 0.5 mA:

$$i_{Tset} = 0.5 \text{ mA},$$

$$R_{set} = \frac{0.0677}{0.5} = 135.4 \, \Omega.$$

Ah, that seems more reasonable. Now I'll choose a more standard value:

$$R_{set} = 150 \, \Omega \, @1\%,$$

$$i_{Tset} = \frac{0.0677}{0.150} = 0.45133 \text{ mA}.$$

The graph of Fig. 1.34 shows what we need to find, the maximum and minimum currents that correspond to the maximum and minimum design temperatures.

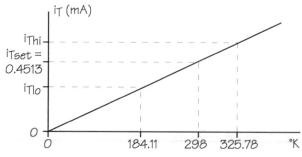

FIGURE 1.34: Temperature–current calibration.

$$i_{Thi} = 0.45133\frac{325.78}{298} = 0.49341 \text{ mA},$$

$$i_{Tlo} = 0.45133\frac{184.11}{298} = 0.27884 \text{ mA}.$$

Now we know what our sensor is going to produce. We also know that we can model it as a current source with an internal resistance of 150 Ω as shown in Fig. 1.35.

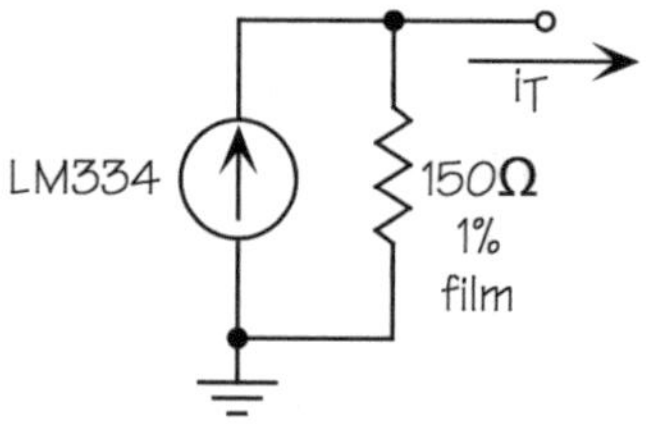

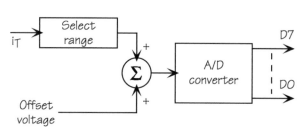

FIGURE 1.35: Current source.

1.6.2 Translation

The next job is to translate the current, which is proportional to temperature, into a range of voltages so that the lowest design temperature gives an output of 0 V and the highest gives an output of +5 V. The general circuit will involve an op-amp to provide range adjustment and offset. Fig. 1.36 shows the general idea.

FIGURE 1.36: Translation circuit.

The circle is a summer that adds the two inputs: one comes from a circuit that converts the current to a voltage, the other is a fixed offset voltage.

I'll do the conversion from current to voltage first. The circuit of Fig. 1.37 uses an op-amp to do this.

How does this op-amp circuit work? Note that the current entering the – node from the left is i_T, while the current entering over the top is $(v_T - 0) / R_f$. No current leaves to enter the – input itself. So the output is found via

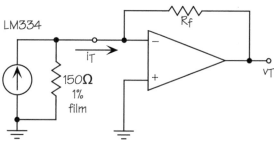

FIGURE 1.37: Conversion of i_T to voltage.

$$i_T + \frac{v_T}{R_f} = 0,$$

$$v_T = -R_f i_T.$$

I will choose R_f by making the range of input currents map to a range of 5 V:

$$(i_{Thi} - i_{Tlo})R_f = 5,$$

$$R_f = 23.30 \text{ k}\Omega.$$

But this must be adjustable. Moreover, the adjustment should use a standard pot and standard resistor values. One single pot does not do it alone. One way would be to put a 1-kΩ pot in series with a 22.5-kΩ resistor, but there is no such resistor. I tried the combinations shown in Fig. 1.38 before choosing the third one in the drawing.

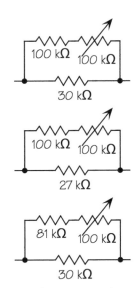

All three use standard resistors and a standard 100-kΩ pot. The first has a range of 23.1 to 26.1 kΩ, the second from 21.3 to 23.7 kΩ, and the third from 21.9 to 25.7 kΩ. I chose the third because it provides a wider, better-centered range of adjustment.

Using R_f = 23.30 kΩ, I can now find the range of voltages that the temperature range will produce:

$$v_{Thi} = -23.30 \times 0.49341 = -11.496 \text{ V},$$

$$v_{Tlo} = -23.30 \times 0.27884 = -6.497 \text{ V},$$

$$range = -11.496 - 6.497 = -4.999 \text{ V}.$$

FIGURE 1.38:   Current-to-voltage adjustment.

Note that I get the right range (5 V), although the range is upside-down because of the inverter. The next op-amp will also invert, so things will be rightside-up after that stage.

The second stage needs to have a summing input that adds 6.497 V to v_T and then inverts the result with unity gain. The block diagram for this stage is shown in Fig. 1.39 and the circuit in Fig. 1.40.

I chose +15 V for the offset input to the summer because that is the likely power-supply voltage for the op-amp. I chose two 10-kΩ resistors for the signal input and feedback resistors because they give unity gain.

But why 10 kΩ? Why not two 1-MΩ resistors? Or some other value? As a general rule, resistors in op-amp circuits should be in the range from about 1 kΩ to about 100 kΩ. Recall that we have made some assumptions to get the ideal op-amp: tiny input currents and tiny input voltages. Resistors that are too large may themselves involve very small currents; resistors that are too small may themselves involve very small voltages. In either case we are tending to make our assumptions less valid.

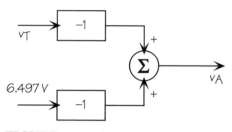

FIGURE 1.39: Sum and invert.

What's the value of R_2? This circuit is an inverter with two inputs, so the output is given by

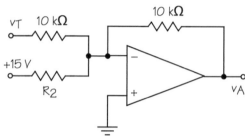

FIGURE 1.40: Circuit for sum and translate.

$$v_A = -\frac{10}{10}v_T - 15\frac{10}{R_2},$$

$$15\frac{10}{R_2} = 6.497 \ for \ v_A = 0,$$

$$R_2 = 23.1\,\mathrm{k\Omega}.$$

Interesting! For R_2 I can use the same two-resistors-and-a-pot circuit that I designed before. That will give me some adjustment of the offset voltage to allow for power-supply and component variations.

1.6.3 Power Supplies

The logic requires a +5-V supply. Op-amps can be run with a single supply, but not conveniently. It is easier to supply both +15 V (to ground) and −15 V (also to ground). This also means that the two op-amps are operating away from their saturation limits. The most negative v_T is about −11.5 V and the most positive v_A is +5 V, both well away from saturation.

LM 334 needs a power supply of from +2 to +25 volts, so +15V will work there, too. However, this voltage and the +15 V supply for the offset voltage probably should be stabilized, which is beyond this course!

1.6.4 System Integration

Fig. 1.41 is the final circuit with everything connected.

I have added the bias compensation resistors. The minus input to the first stage is dominated by the 150-Ω resistance of the current source, so I used 150 Ω for bias compensation. The

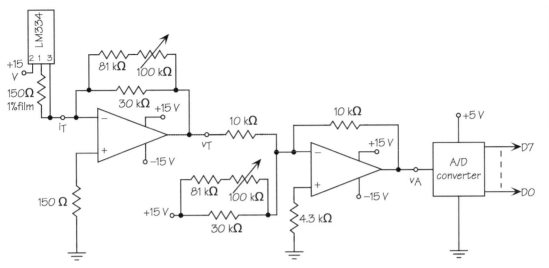

FIGURE 1.41:   Bank thermometer: analog portion.

minus input to the second stage "sees" three parallel paths: 10 kΩ, 23.1 kΩ (the lower group of three), and 10 kΩ. In parallel, these are about 4.1 kΩ, so I chose the nearest commercial 5% value of 4.3 kΩ. (Bias compensation does not need to be exact.)

1.7   SUMMARY

The purpose of this chapter has primarily been to review things you probably already knew:

- Fundamentals, including Ohm's law for resistors, capacitors, and inductors, Kirchhoff's current law, Kirchhoff's voltage law, and independent and dependent sources.
- Some simple but useful tools: series and parallel resistance, voltage dividers and pots, and current dividers.
- The "big gun" approach to writing equations for DC circuits, circuits in the time domain, and circuits operating in the sinusoidal steady state, including impedance.
- Op-amps and some simple applications of them.
- Finally, a complete design problem that we can handle with what we already know.

If some of this seems unfamiliar, this would be a good time to review! These topics form the foundation of just about everything we'll be doing in the rest of the course.

What's important in all of this? I would be very happy if you could do just one thing with 100% reliability: write a set of equations that describes a given circuit.

What's next? It's one thing to be able to write those equations. It's another to do this efficiently. Even with computers, we are better off with fewer equations. On top of that, we need ways of gaining insight about circuit configurations without writing lots of equations. Or perhaps without writing any at all.

So in the next chapter we will look at more formal ways of approaching circuits. In a later chapter we will see some more specialized techniques that aid our understanding of how many different circuits work.

CHAPTER 2

Formal Circuit Analysis: Big Gun = Hard Way

Right now, our primary technique for analyzing circuits is the "big gun." We write element constraints for all the devices, Kirchhoff's current law at all but one node, and Kirchhoff's voltage law around all the meshes. This technique gives us a whole pile of equations.

A not-too-uncommon circuit might be a passive filter. The analysis of that filter might involve seven circuit elements. From what we know now, this analysis is going to require 14 equations (two times the number of elements). Our goal in this chapter is to reduce that number.

Reduce it we can! The filter that yields 14 equations can generally be done with no more than 3! Not a bad reduction, I think you'll agree. But to do this, we need a little better understanding of the organization of a circuit and what we can choose to eliminate. 14:3? It should be worth the effort!

2.1 CIRCUIT TOPOLOGY

Circuit topology can be a very formal mathematical topic. One reason is that the mathematician wants to visualize all sorts of different—and strange—shapes of circuits. But practical circuits don't come in esoteric shapes, so we can ignore the fancy stuff and concentrate on the basics.

What I'm going to do here is present some definitions and then just wave my hands and argue that some things are fairly obvious. That'll keep us out of the clutches of formal mathematics and, I think, give us a better understanding of what's really happening.

2.1.1 What is a Circuit?

Hmmm, that's easy. A circuit is a bunch of electrical elements hooked together with conductors in some useful way. Fig. 2.1 is such a circuit. Note that current can flow in the circuit because there are closed paths—that's

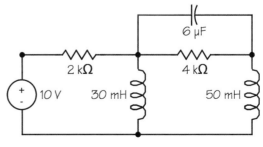

FIGURE 2.1: A circuit.

what I mean by "useful way." The circuit has a source in it and some passive elements that presumably do something to the currents and voltages.

This circuit has six elements, so an analysis using the "big gun" requires 12 equations. But let's look at some other properties of this circuit.

First, we concentrate all of the properties of each element at that element. To say this another way, we *lump* the element's properties into the symbol.

Second, we ignore the conductors. They are perfect, which means they have no resistance. Now, there are some cases where this is not true, so to get out of that situation, we lump the conductor resistance into the element.

Third, we ignore the time electrons take to get through our elements and conductors. In other words, our elements are small enough that time is not a factor. Moreover, our conductors are so perfect that transit times in them are zero, too.

Now let's define four things that will be the heart of our definitions as we reduce the number of equations:

Node—a junction where two or more elements meet.

Branch—a path between two nodes, generally through a circuit element.

Loop—any sensible closed path. (A less-than-sensible closed path would be the one that passes through some branch more than once.)

Mesh—a loop that doesn't have any loops inside it.

Now let's use these in our quest for the reduced set of equations.

2.1.2 Some Simple Topology

When we study just the topology of a circuit, we can ignore the elements themselves. We are interested just in the paths and their connections. The circuit that we saw in Fig. 2.1 can be reduced to a set of lines and connections as shown in Fig. 2.2.

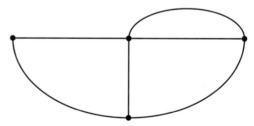

FIGURE 2.2: Graph of "A circuit".

The drawing is called a *graph*, which reduces the circuit to just its branches and nodes. These will lead us to some conclusions without having elements getting in our way.

A *tree* includes all the branches of the graph that you can draw without making any loops. Let's start with just the nodes (Fig. 2.3).

Now add one branch arbitrarily (Fig. 2.4).

FIGURE 2.3: Nodes of our graph.

Add a second branch, almost arbitrarily, making sure it creates no loop (Fig. 2.5).

Add a third branch, again being careful not to add one that closes a loop (Fig. 2.6).

Try to add another branch! You can't do it because there are no other branches that can be included in our graph without closing a loop somewhere. What we have in Fig. 2.6 is a *tree* for this circuit. This is not the only tree we can make, but every tree for this graph will have exactly three branches.

Why exactly three? There are four nodes. The first branch included two of them in the tree. Each additional tree branch adds one more node to the tree. Here we have four nodes, so the tree must have three branches. If N is the number of nodes and T is the number of tree branches, then

$$T = N - 1.$$

Since these tree branches connect every node, I can assign voltages to the tree branches and include every node in my definitions. The tree graph in Fig. 2.7 shows some assigned voltages. I chose signs arbitrarily.

These voltages form a *complete set* because from them I can define any other voltage in the circuit. For example, the voltage between the two nodes at the top left (+ on the left) is found from Kirchhoff's voltage law around the loop that includes v_1 and v_3:

$$-v_1 + v_{left} + v_3 = 0,$$

$$v_{left} = v_1 - v_3.$$

Similarly, the voltage between the right end and the bottom (+ on the right) is found in terms of v_3 and v_4:

$$-v_3 + v_4 + v_{right} = 0,$$

$$v_{right} = v_3 - v_4.$$

FIGURE 2.4:   One branch of a tree.

FIGURE 2.5:   Two branches of a tree.

FIGURE 2.6:   Three branches of a tree.

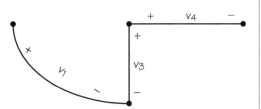

FIGURE 2.7:   Tree voltages.

We have the voltage across every element of the original circuit defined, either as a tree-branch voltage or a simple combination of tree-branch voltages. So we conclude that we can define all voltages in terms of the T tree-branch voltages. That's why I said that these tree-branch voltages form a complete set.

To be a little more pure, I also need to note that you can't make this complete set with less than T voltages. That's because the tree would be incomplete because it would not include all the nodes.

Am I done yet? Nope! There's another part to the story. Let's define a chord as a branch that closes a loop in our graph. Start with the tree and add one more branch. In Fig. 2.8, I've added the branch between the two nodes on the left (it's shown dotted).

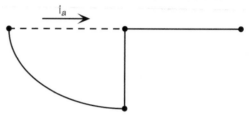

FIGURE 2.8:   Tree and one chord.

How many chords are there? There were T tree branches. All the rest of the branches are chords because they close loops. So if C is the number of chords possible and B is the total number of branches, then

$$C = B - T = B - (N - 1).$$

That chord closes a loop, so a current could now flow in the branches of our graph. I've labeled that current i_a. But since this is the only loop, this current i_a flows in the chord and the two adjacent tree branches. We commonly label that current in a slightly different way as shown in Fig. 2.9.

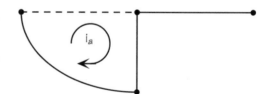

FIGURE 2.9:   Relabeled loop current.

The circular arrow in the drawing indicates that the current i_a flows through the chord and two tree branches. The arrow goes clockwise because it doesn't go counterclockwise. Oh, really? Well, you have to choose a direction, and I generally choose the clockwise.

Now let's put all the remaining branches into our graph. Each is a chord and each closes a loop. So I can label the loop currents as shown in Fig. 2.10.

The next argument is about the same as we used for voltages. The three loop currents shown form a complete set. Why? Each

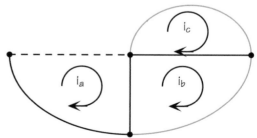

FIGURE 2.10:   Complete graph with currents.

time we added a chord, we defined a new loop current. We have added all the chords and hence have three loop currents.

But that isn't enough. First, we generally don't care, in our final result, what the loop currents are. Instead, we are more likely to want currents in certain branches. These can be derived from the loop currents.

For example, the current in the upper left branch is i_a, positive to the right. The current in the vertical middle branch can be found from the loop currents. If we define this current as positive downward, then

$$i_{middle} = i_a - i_b.$$

Why? Because i_a is a current in that branch that is caused by placing the upper left chord into the graph, and i_b is the current caused by placing the lower right chord into the graph. The current i_a is in the same direction as our new current i_{middle} so it is plus, while i_b is in the opposite direction, so it is minus.

We have the current through every element of the original circuit defined, either as a loop current (defined by a chord) or a simple combination of loop currents. So we conclude that we can define all currents in terms of the C chord currents. That's why I said that these chord currents form a complete set.

Where has this gotten us? If a circuit has B branches and N nodes, then

- there are T = N − 1 tree branches and hence T voltages in the complete voltage set, and
- there are C = B − T = B −(N − 1) chords and hence C currents in the complete current set.

We'll see where this all takes us in the next section.

2.1.3   Number of Equations

It's crunch time! Time to show that all this gets us somewhere. I'm going to demonstrate that we can write a complete set of equations for a circuit in either of two ways:

- write $T = N − 1$ equations using Kirchhoff's current law, or
- write $C = B − T = B −(N − 1)$ equations using Kirchhoff's voltage law.

Either of these sets of equations will be a complete set that can be used to find everything about the circuit. We can write and solve either set and get the same results. (If T and C are much different, it might be smart to choose the smaller one!)

Let's follow up the first of the two ways stated above. The circuit of Fig. 2.11 has every voltage and current variable labeled (as if we were starting to fire the big gun).

This circuit has six branches, so we expect to write, when we fire the big gun, 12 equations in 12 unknowns. But half of those are element constraints, generally Ohm's law or source values.

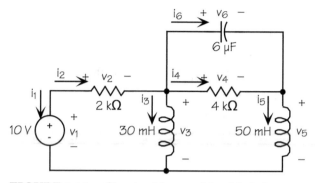

FIGURE 2.11: Circuit with everything labeled.

In the general case, our circuit of B branches would yield $2B$ equations. But Ohm's law and source values would reduce this number to B, so we would have only B equations with B unknowns left.

Now consider the T tree branches. T must be smaller than B because not all branches can be tree branches. If we define all B branch voltages, then T of them form a complete set and the remaining $B–T$ of them can be derived using KVL as I've shown. So there are T independent voltages. Why not use them as the variables? Then there are only T equations to be solved.

In our example, there are only three tree branches. Hence we need only three equations using the tree-branch voltages. Which would you rather solve, the 12 equations of the big gun or the 3 of this method?

I'm going to skip the parallel argument for the chords, at least for a while, and go on to the equations themselves. Remember that, by using tree-branch voltages as our variables, we can reduce significantly the number of equations we need to solve. What we'll need is a systematic way of selecting the tree-branch voltages and writing the appropriate equations.

2.2 NODAL ANALYSIS

Nodal analysis is a systematic way of selecting the tree-branch voltages and then writing the equations. We will be selecting T voltages and then writing T equations. An example is probably the easiest way to start.

2.2.1 Example I

The circuit shown in Fig. 2.12 has five elements. The big-gun approach would have us defining five voltages and five currents, then writing ten equations to get the result.

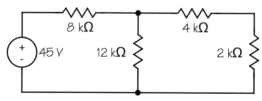

FIGURE 2.12: Circuit for Example I.

The goal for this circuit is to find the voltage across the 12-kΩ resistor. My first step will be to define a tree, but I'm going to choose its branches in a special way. My tree is shown in Fig. 2.13, where I have chosen *all* the tree branches to have a *common* reference point.

I have marked this reference node with the ground symbol to indicate that it is the reference node. We can also call this node the *ground node* or *the datum node*.

Then I define all the tree-branch voltages with their *minus* signs at the reference node. That gives me three voltages, v_1, v_2, and v_3, that form the complete set of voltage variables.

But the graph is not a convenient place to show these, especially because we need to work with the circuit elements as well. So on the original circuit I will show these voltages as I have done in Fig. 2.14.

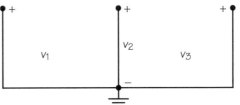

FIGURE 2.13: Tree for Example I.

The three voltages are labeled without putting their minus signs on the reference node. We call these three voltages *node voltages* rather than referring to the original tree branches. Hence the method of analysis that we are following is called *nodal analysis* because it uses these node voltages as its set of variables.

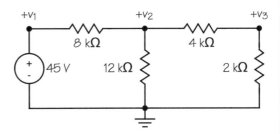

FIGURE 2.14: Labeled voltages.

OK, now let's see how to write the three equations. One is pretty obvious, because the voltage v_1 is along the branch that holds the 45-V source. So our first of three equations is

$$v_1 = 45.$$

The other two equations come from the application of Kirchhoff's *current* law at the other two labeled nodes. Note that we are going to write *current* equations, even though this seems strange, having just defined voltages. But since we know, in terms of the node voltages, the voltage across each element, we can write an expression for the current through that element.

I am going to choose currents *leaving* at a node to be positive. So my equations will have negative signs only if I define a current as arriving.

Look at the node labeled v_2. There are three currents involved. Each can be found from the voltage across its element divided by the resistance. If I work in milliamperes and kilohms, the three currents are

- to the left, $(v_2 - v_1) / 8$,
- down, $v_2 / 12$, and
- to the right, $(v_2 - v_3) / 4$.

Combining these into a KCL equation for the node v_2, I get

$$\frac{v_2 - v_1}{8} + \frac{v_2}{12} + \frac{v_2 - v_3}{4} = 0.$$

Finally, consider the node labeled v_3. Two currents are involved:

- to the left, $(v_3 - v_2) / 4$, and
- down, $v_3 / 2$.

The third equation is the combination of these two terms, again paying attention to sign:

$$\frac{v_3 - v_2}{4} + \frac{v_3}{2} = 0.$$

Now I have three equations in three unknowns. Don't miss the fact that I wrote them by writing expressions for currents leaving the nodes. In each case, the current is the voltage difference across the element divided by the element value.

Here are all three equations, along with the solution for the voltage across the 12-kΩ resistor (v_2):

$$v_1 = 45,$$

$$\frac{v_2 - v_1}{8} + \frac{v_2}{12} + \frac{v_2 - v_3}{4} = 0,$$

$$\frac{v_3 - v_2}{4} + \frac{v_3}{2} = 0,$$

$$\text{Solution} \qquad v_2 = 15 \text{ volts.}$$

Was that easier than writing and solving ten equations in ten unknowns? Maybe not, because this method requires more thinking. But it's a powerful method that is often used by computer programs that simulate circuits (SPICE is an example).

2.2.2 Steps for Nodal Analysis

It's probably convenient here to list the steps for analyzing a circuit using the nodal-analysis technique. They are rather easily reduced to a straightforward algorithm:

1. Select and label a reference node. Usually there is an obvious reference node at the bottom of the circuit. But the selection makes no difference in the rest of the algorithm.
2. Label all the remaining nodes with voltage variables. Note that no mention is made of the tree branches, but that is what we are doing. However, at times (as in the next example) we will define a node voltage that does not use a tree branch that exactly follows one of the real branches in the circuit.
3. Write a KCL equation for each of the nodes, using the node voltages as the variables. Choose leaving currents to be positive.
4. Solve these equations by any method that works for you.
5. Answer the questions posed about the circuit, since we often want more than just the node voltages.

Now let's use this algorithm to solve another problem.

2.2.3   Example II

The circuit of Fig. 2.15 is a circuit for which the value of the output voltage v_o is wanted. The circuit contains a dependent source, but this does not complicate the equations.

Step 1 of the algorithm says to label a reference node, which I am going to choose at the bottom of the circuit. Step 2 says to label the other nodes with node-voltage variables. Fig. 2.16 shows the results of these steps.

Step 3 says to write KCL at each of the nodes. But at two of these nodes we already know the voltage. At v_1 the voltage is 8 V; at v_3 the voltage is $4000i_x$. We don't write KCL equations for these because we don't know the currents through the voltage sources. That leaves the node labeled v_2. Three currents are involved:

- to the left, $(v_2 - v_1) / 2000$,
- down, $(v_2 - v_3) / 500$, and
- to the right, i_x.

Because of the dependent source, a constraint equation is also required, since i_x is another

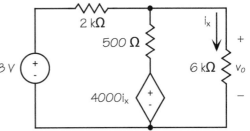

FIGURE 2.15:   Circuit for Example II.

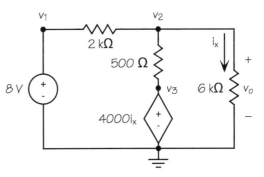

FIGURE 2.16:   Labeled circuit for Example II.

variable. We get that from the 6-kΩ resistor: $i_x = v_2 / 6000$. (Note that v_2 and v_o are the same voltage; we could have saved some writing by using v_o as the node voltage in place of v_2.)

Here are the three equations plus the constraint:

$$v_1 = 8,$$

$$\frac{v_2 - v_1}{2000} + \frac{v_2 - v_3}{500} + i_x = 0,$$

$$v_3 = 4000 i_x,$$

$$i_x = \frac{v_2}{6000}.$$

Step 4 says to solve this set, which yields $v_1 = 8$ V, $v_2 = 3$ V, $v_3 = 2$ V, and $i_x = 0.5$ mA.

Step 5 says to answer the question, so we look back to find that the goal was to get v_o. Since this is the same voltage as v_2, the solution is

$$v_o = v_2 = 3 \text{ volts.}$$

Why did I write the equations in amperes and ohms rather than a somewhat simpler milliamperes and kilohms? The dependent source messes me up! In fact, the units of the "4000" on that source are volts per ampere. I'd get confused if I tried to change it!

2.2.4   Example III

Consider a little more complicated circuit that has a current source (see Fig. 2.17). I've already labeled the reference node (Step 1) and the node voltages (Step 2). Our goal is to find v_o.

Step 3 is not really complicated by the current source, because the 5-mA source is

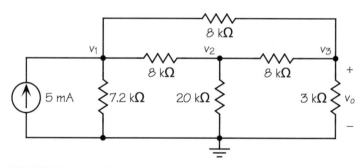

FIGURE 2.17:   Circuit for Example III.

merely delivering 5 mA to the node labeled v_1. This time I'll just write the complete set of equations; you should check through them so that you are sure where the terms come from. (I've written in milliamperes and kilohms.)

$$-5+\frac{v_1}{7.2}+\frac{v_1-v_2}{8}+\frac{v_1-v_3}{8}=0,$$

$$\frac{v_2-v_1}{8}+\frac{v_2}{20}+\frac{v_2-v_3}{8}=0,$$

$$\frac{v_3-v_1}{8}+\frac{v_3-v_2}{8}+\frac{v_3}{3}=0.$$

Step 4 says to solve these, which yields 18, 10, and 6 volts for the three nodes (left to right). Step 5 says to answer the question, in our case to find v_o, which is the same as v_3:

$$v_o = v_3 = 6 \text{ volts.}$$

2.2.5 Example IV

Here's a deceptively simple problem (Fig. 2.18) that has a twist to it. The circuit in the drawing is already labeled according to Steps 1 and 2. The goal is to find v_o.

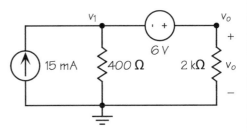

FIGURE 2.18:   Circuit for Example IV.

Step 3 wants us to write the KCL equations for the nodes. But how many? If I start writing the KCL equation for the node labeled v_1, I note that there is a 15-mA current arriving, a current of $v_1/400$ leaving, and a current of…. Hmmmm, what is the current through the 6-V source?

This situation complicates the node equation because we can't find the current through the voltage source by looking at the voltage source. Where, then? Look at the 2-kΩ resistor. The current down through it is $v_o/2000$. That is the same current that is leaving the node v_1 and going through the voltage source.

Aha! So I *can* write the KCL equation for v_1. But I've "used up" v_o and I still need another equation. That will come from the fact that the 6-V source establishes the voltage difference between v_o and v_1. So the equation $v_o - v_1 = 6$ becomes my second equation.

Here are the two equations for this analysis (written in mA and kΩ), along with the solution for v_o:

$$-15+\frac{v_1}{0.4}+\frac{v_o}{2}=0,$$

$$v_o - v_1 = 6,$$

$$\text{Solution} \qquad v_o = 10 \text{ V.}$$

A voltage source in a position like the 6-V source will always establish the difference between two node voltages. Then we have to look beyond the source to find the current through it.

2.2.6 Example V

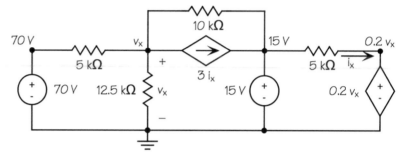

FIGURE 2.19:   Circuit for Example V.

Fig. 2.19 will end all this with a problem that embodies several sources. The goal is to find the power supplied by each of the four sources.

I've labeled the reference node and the node voltages. Note that I have taken advantage of the known voltages at two nodes relative to the reference, +70 V and +15 V.

The surprise is that there is only one unknown node voltage! Two other nodes are fixed by independent sources, and a third is fixed by a dependent source. We do need a constraint equation for i_x, however, so there will be two equations.

The first equation is KCL at the v_x node; the second is the constraint for i_x.

$$\frac{v_x - 70}{5} + \frac{v_x}{12.5} + 3i_x + \frac{v_x - 15}{10} = 0,$$

$$i_x = \frac{15 - 0.2v_x}{5}.$$

Solution $v_x = 25 \text{ V}, i_x = 2 \text{ mA}.$

The problem asked for the power supplied by each source, so we need to do Step 5 and answer the question asked. Each of the four is a little different problem:

70-Volt source: We know the voltage is 70 V. We need the current flowing upward through the source. To get this, let's find the current to the right through the 5-kΩ resistor. That current is $(70 - v_x) / 5$ mA, so the power supplied by the 70-V source is

$$P_{70} = 70 \times \frac{70 - v_x}{5} = 630 \text{ mW}.$$

Dependent current source: This time we know the current through the source. Now we need the voltage across it, positive at the right end. This voltage is $(15 - v_x)$ so the power supplied is

$$P_{3ix} = (15 - v_x) \times (3i_x) = -60 \text{ mW}.$$

15-Volt source: Here we know the voltage. It's the current flowing upward that is hard to get. Note that there are three branches, and we can find the current leaving the upper node through each of them separately. The resultant equation is

$$P_{15} = 15 \times \left[\frac{15 - v_x}{10} - 3i_x + i_x \right] = -75 \text{ mW}.$$

Dependent voltage source: Again, the voltage is known and the current upward through the source is $-i_x$. The result is

$$P_{02vx} = (0.2v_x) \times (-i_x) = -10 \text{ mW}.$$

Total power supplied: The total power supplied by the four sources is the sum of these results: 485 mW. It is left as an exercise for the student (arggghhh!) to find the power absorbed by each resistor and see if the result is 485 mW.

That's nodal analysis. Don't forget the basic steps:

1. Define the reference node.
2. Label the node-voltage variables.
3. Write KCL equations at each node except the reference, plus any constraint equations needed.
4. Solve the equations.
5. Answer the questions posed.

In the next section we are going to do a very similar thing, but this time with chords and currents and KVL.

2.3   MESH ANALYSIS

Remember chords? No, not G⁷, although that might be fun. Chords were the branches that were left over when we made a tree graph of our network. Since there are $T = N - 1$ branches (where N is the number of nodes), the number of chords must be $C = B - T = B - (N-1)$, where B is the number of branches.

Remember also, please, that installing a chord in the tree closes one loop and leads to one current. We can define these loop currents by installing chords one at a time. All other currents in the circuit can be found from these loop currents.

The result of all this is that there are C chords and C loop currents that form an independent set of currents, and finally, C equations in C unknowns. The process by which we get to these equations is called *mesh analysis*.

A *mesh* is a loop that has no loops inside it. But a problem arises when the circuit is not planar, that is, when it cannot be drawn on a plane[1] without crossovers. Meshes get impossible to define. So we restrict mesh analysis to planar circuits. This is not much of a restriction, though, because many of our practical circuits are planar.

Let's do this by example, starting with a simple circuit and showing the chords. After that, we'll develop a simple way of choosing the currents without thinking about chords.

2.3.1 Example VI

Fig. 2.20 shows the same circuit that we started nodal analysis with. Since our interest is in the loops, and in fact, the meshes, it's pretty obvious that there are two meshes and a path around the outside.

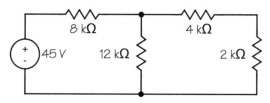

FIGURE 2.20: Circuit for Example VI.

Recall that we wanted the voltage across the 12-kΩ resistor. To start the analysis, I'm going to draw a possible tree for this network; it's in Fig. 2.21.

I've also drawn the two chords, each of which gives rise to one loop current. To be correct, I should call these currents *mesh currents* because they are defined in the meshes of the circuit. Now let's combine the graph and the circuit in Fig. 2.22.

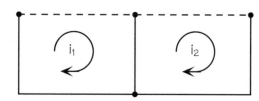

FIGURE 2.21: Tree for Example VI.

The next step is a straightforward application of Kirchhoff's voltage law. I'll apply it twice, once around each of the meshes. I'm also going to be careful of the directions by setting for myself two rules:

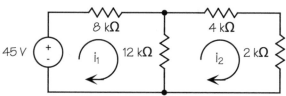

FIGURE 2.22: Example VI with mesh currents.

- Go around the mesh clockwise to sum voltages.
- When encountering a voltage, take the first sign as the sign of the voltage in the equation.

[1] Technically, things are O.K. if the circuit can be drawn on a sphere.

But what voltages? Consider the 8-kΩ resistor at the top. The voltage across that resistor can be defined in terms of i_1. Its magnitude is $8i_1$ (in milliamperes). The plus end of the voltage will be on the left.

How about the voltage across the 12-kΩ resistor? I'll choose plus to be at the top, which says that the positive current flowing through the resistor is $i_1 - i_2$. In other words, the current flowing down through that resistor is $i_1 - i_2$ and the voltage will have the plus sign at the top.

Now let's write KVL around the loop according to my rules for KVL:

$$-45 + 8i_1 + 12(i_1 - i_2) = 0.$$

I started in the lower left corner and moved clockwise. I first encountered the 45-V source and ran into its minus sign first. So the first term is −45. Then I got to the 8-kΩ resistor. Since i_1 is flowing to the right, the passive sign convention says the plus sign is on the left. So the voltage term is $8i_1$. Finally, I encountered the 12-kΩ resistor. Again, the plus sign is at the top because I'm taking the direction of i_1 as defining the passive sign convention. But i_2 is flowing the other way. Hence the last term has a minus in the middle. And we are back to the beginning.

How about the second mesh? I'll start in the lower left corner and go clockwise. Since I am following i_2, I'll let it define "positive" for me. So the first voltage is across the 12-kΩ resistor, plus at the bottom as I go around this second mesh. That makes the first term of the equation $12(i_2 - i_1)$. The other two terms are easier because they involve only the current i_2 and the individual resistors. The result is

$$12(i_2 - i_1) + 4i_2 + 2i_2 = 0.$$

Solving these gives $i_1 = 3.75$ mA, $i_2 = 2.5$ mA. But that doesn't answer the question of what is the voltage across the 12-kΩ resistor. We know the currents, though, so that the voltage is

$$v_{12} = 12(i_1 - i_2) = 15 \text{ V}.$$

Wow! That's what we got before using nodal analysis.

2.3.2 Steps for Mesh Analysis

Let's list the steps for analyzing a circuit using the mesh-analysis technique. They are rather easily reduced to a straightforward algorithm:

2. Draw a current variable in each mesh. Generally choose clockwise currents for consistency.

3. Write a KVL equation for each of the meshes, using the mesh currents as the variables. Choose the first sign of the voltage as you come to it as the sign of that voltage.
4. Solve these equations by any method that works for you.
5. Answer the questions posed about the circuit, since we often want more than just the mesh currents.

Hmmm, did I mistype that? Where's Step 1? For the moment, just start at Step 2 and solve another problem.

2.3.3 Example VII

The circuit of Fig. 2.23 already has Step 2 done to it. There are two mesh currents, but note that i_2 could have been called "i_x" because they are both the same.

Step 3 has me going around each of the two meshes and writing the KVL relationship. In each case I'll start in the lower left corner. For the left-hand mesh, there will be four terms:

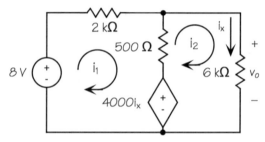

FIGURE 2.23:   Circuit for Example VII.

- −8 from the source, where I use the minus sign because that's the one I come to first;
- $2000i_1$, chosen to be plus because i_1 is entering from the same end I am;
- $500(i_1 - i_2)$ because the current in *my* direction through the 500-Ω resistor is i_1 in my direction and i_2 in the opposite direction; and
- $4000i_x$ with a plus sign.

Track around the right-hand mesh, please, and then check my equations:

$$-8 + 2000i_1 + 500(i_1 - i_2) + 4000i_x = 0,$$
$$-4000i_x + 500(i_2 - i_1) + 6000i_2 = 0,$$
$$i_x = i_2.$$

Oh, where did that last equation come from? Because there is a dependent source in the circuit, we need a constraint equation.

Step 4 says to solve 'em, so I did, with the result that $i_1 = 2.5$ mA, $i_2 = i_x = 0.5$ mA.

Step 5 wants the answer to the question, which here was to find v_o:

$$v_o = 6000i_x = 3 \text{ volts.}$$

Gosh, that's what we got before using nodal analysis.

2.3.4   Example VIII

Fig. 2.24 is a rather ordinary circuit with the mesh currents already shown. The source, though, is a current source that establishes the current i_1 in the leftmost mesh.

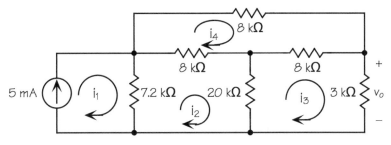

FIGURE 2.24:   Circuit for Example VIII.

The mesh equations (Step 3) for this circuit are straightforward. Note that i_1 is constrained by the 5-mA source. I've written the equations in kΩ and mA:

$$i_1 = 5,$$
$$7.2(i_2 - i_1) + 8(i_2 - i_4) + 20(i_2 - i_3) = 0,$$
$$20(i_3 - i_2) + 8(i_3 - i_4) + 3i_3 = 0,$$
$$8(i_4 - i_2) + 8i_4 + 8(i_4 - i_3) = 0.$$

The Step 4 solution yields i_1 = 5, i_2 = 2.5, i_3 = 2, and i_4 = 1.5, all in milliamperes. Finally, Step 5 reminds us to find v_o:

$$v_o = 3i_3 = 6 \text{ volts.}$$

2.3.5   Example IX

The circuit shown in Fig. 2.25 gave us some trouble when we were writing node equations, if you'll recall. This was caused by the 6-V source that made it harder to find the current flowing through it. That problem doesn't arise here.

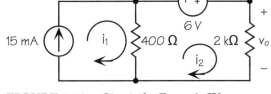

FIGURE 2.25:   Circuit for Example IX.

I'm going to combine Step 3 and Step 5 this time by writing the equation for v_o at the same time that I write the KVL equations:

$$i_1 = 15,$$
$$0.4(i_2 - i_1) - 6 + 2i_2 = 0,$$
$$v_o = 2i_2.$$

The solution to these yields v_o = 10 V as before.

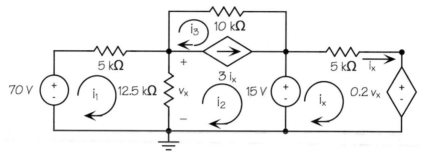

FIGURE 2.26: Circuit for Example X.

2.3.6 Example X

One more example! In the circuit shown in Fig. 2.26, we are to find the power supplied by each of the sources.

The equations are straightforward—almost. The trouble comes in the i_2 and i_3 meshes. Consider writing KVL starting in the lower left corner of i_2:

- The first term is easy: $12.5\,(i_2 - i_1)$.
- And what's the voltage across the dependent current source? We can't find it directly, so instead of continuing around the i_2 path, we go up over the top. (My path now includes the 12.5-kΩ resistor, the 10-kΩ resistor, and the 15-V source.) That term is $10\,i_3$.
- Coming back down to the i_2 mesh, we encounter the 15-V source that yields the term +15.

Now there's a new problem. We have used two meshes together and written just one equation. Yet there are two variables, i_2 and i_3. We need another equation. That's provided by the dependent source itself. Note that the current through it is constrained by the source to be $3i_x$. In terms of our mesh equations, the current through it (in the same direction) is $i_2 - i_3$. There's the needed equation: $i_2 - i_3 = 3i_x$.

Here's the full set of equations, including the constraint equation for v_x. (I avoided the need for a constraint equation for i_x by using i_x as the mesh current in the rightmost mesh.)

$$-70 + 5i_1 + 12.5(i_1 - i_2) = 0,$$
$$12.5(i_2 - i_1) + 10i_3 + 15 = 0,$$
$$-15 + 5i_x + 0.2v_x = 0,$$
$$i_2 - i_3 = 3i_x,$$
$$v_x = 12.5(i_1 - i_2).$$

The solution to this set is i_1 = 9 mA, i_2 = 7 mA, i_3 =1 mA, i_x =2 mA, and v_x = 25 V. Now let's answer the power question again:

70-V source: The current upward through the source is i_1, so the power supplied is

$$P_{70} = 70 \times i_1 = 630 \text{ mW}.$$

Dependent current source: The current through this source is $3i_x$, but the voltage must be found from something else. Let's find it by calculating the voltage across the 10-kΩ resistor, keeping the plus sign to the right so that we get the power supplied by the source. The voltage across the 10-kΩ resistor is $10\,(-i_3)$, so the power supplied by the source is

$$P_{3ix} = (-10i_3) \times 3i_x = -60 \text{ mW}.$$

15-V source: The current flowing up through this source is $i_x - i_2$, which yields for the power supplied

$$P_{15} = 15(i_x - i_2) = -75 \text{ mW}.$$

Dependent voltage source: The voltage is $0.2\,v_x$ and the current is $-i_x$. That makes the power supplied

$$P_{02vx} = (0.2v_x) \times (-i_x) = -10 \text{ mW}.$$

These are the same results we got using nodal analysis.

That's mesh analysis. Don't forget the basic steps:

2. Label the mesh-current variables.

3. Write KVL equations for each mesh, plus any constraint equations needed.

4. Solve the equations.

5. Answer the questions posed.

2.4   MORE EXAMPLES

Since this chapter does mostly nodal and mesh analysis, that's what I will do in these examples. The first example is a counting exercise. The rest are pairs of problems: DC, DC with a dependent source, time domain, and phasor domain.

2.4.1   Example XI

How many nodes, branches, tree branches, chords, independent voltages, and independent currents are there in the circuit of Fig. 2.27?

Nodes connect elements, but dots aren't necessarily nodes and nodes don't necessarily have dots on them. In the circuit, the whole bottom line is a node, even though there are two dots on it. The upper left and upper right corners are nodes, even though they don't have dots. Finally both ends of the top voltage source are nodes. The circuit has $N = 5$ nodes.

Branches are elements. The circuit has six resistors and two voltage sources, so it has $B = 8$ branches.

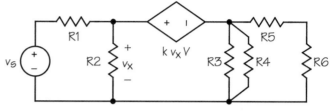

Tree branches are as many branches as can be included in the circuit without making loops: $T = N - 1 = 4$.

FIGURE 2.27: Example XI.

Chords are the remaining branches: $C = B - T = 4$.

The number of independent voltages in the circuit is the same as the number of tree branches, which is 4. (This is also the number of nodes minus 1.)

The number of independent currents is the same as the number of chords, which is 4.

2.4.2 Example XII

Find v_o in the circuit of Fig. 2.28 using nodal analysis.

I will choose the reference node to be at the bottom of the circuit. My node voltage labels are +150 at the top left and v_o at the right. The single node equation is

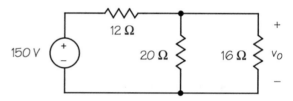

FIGURE 2.28: Examples XII and XIII.

$$\frac{v_o - 150}{12} + \frac{v_o}{20} + \frac{v_o}{16} = 0,$$

which solves to yield

$$v_o = 63.83 \text{ V}.$$

2.4.3 Example XIII

Use mesh analysis to find v_o in the same circuit (Fig. 2.28).

My mesh currents are i_1 in the left mesh and i_2 in the right mesh, both clockwise. These two mesh equations require a third equation to find the value of v_o:

$$-150 + 12i_1 + 20(i_1 - i_2) = 0,$$
$$20(i_2 - i_1) + 16i_2 = 0,$$
$$v_o = 16i_2.$$

The solution is the same as before (fortunately!).

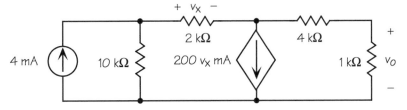

FIGURE 2.29:   Examples XIV and XV.

2.4.4   Example XIV

The circuit of Fig. 2.29 contains a dependent current source. Find v_o.

I will as usual choose the bottom node as the reference, then label the nodes at the top v_1, v_2, and v_o from left to right. The node equations are

$$-4\times10^{-3} + \frac{v_1}{10\times10^3} + \frac{v_1 - v_2}{2\times10^3} = 0,$$

$$\frac{v_2 - v_1}{2\times10^3} + 200\times10^{-3}v_x + \frac{v_2 - v_o}{4\times10^3} = 0,$$

$$v_x = v_1 - v_2,$$

$$\frac{v_o - v_2}{4\times10^3} + \frac{v_o}{1\times10^3} = 0.$$

The result is

$$v_o = 8.048 \text{ V}.$$

2.4.5   Example XV

Find v_o in the same circuit (Fig. 2.29) via mesh analysis.

My currents are i_1, i_2, and i_3 clockwise in the meshes from left to right. But the presence of current sources makes the equations more interesting:

$$i_1 = 4\times10^{-3},$$

$$\frac{i_2 - i_1}{10\times10^3} + 2\times10^3 i_2 + 4\times10^3 i_3 + 1\times10^3 i_3 = 0,$$

$$i_2 - i_3 = 200\times10^{-3}v_x,$$

$$v_x = 2\times10^3 i_2,$$

$$v_o = 1\times10^3 i_3.$$

The first equation is the result of the source on the left, which constrains i_1 to be the value of the source.

The second equation is written around the remainder of the circuit. The dependent current source in the middle constrains the current but not the voltage, so Kirchhoff's voltage law cannot be made into a complete loop through this source.

The third equation takes into account that dependent source by writing that it constrains the current $i_2 - i_3$ through that branch.

The fourth equation defines the voltage constraint v_x.

The last equation finds the required result from the currents. The result is the same as before.

2.4.6 Example XVI

The circuit of Fig. 2.30 has been sitting with the switch open for a long time before t = 0. This means that there can be no stored energy in the circuit because we assume that anything that had been stored long ago has dissipated. Therefore the initial current through the inductor and the initial voltages across the capacitors are all zero.

My labels have the reference node at the bottom, node +24 at the top left, node v_1 in the middle, node v_2 at the top of the dependent source, and node v_o at the top right. The equations involve both integrals and derivatives:

$$\frac{v_1 - 24}{100} + 2 \times 10^{-6} \frac{dv_1}{dt} + \frac{1}{120 \times 10^{-3}} \int_0^t (v_1 - v_o)dx + 0 = 0,$$

$$\frac{1}{120 \times 10^{-3}} \int_0^t (v_o - v_1)dx + 0 + 1 \times 10^{-6} \frac{d(v_o - v_2)}{dt} + \frac{v_o}{2 \times 10^3} = 0,$$

$$v_x = 24 - v_1,$$

$$v_2 = 3v_x,$$

$$v_1(0) = 0,$$

$$v_o(0) - v_2(0) = 0.$$

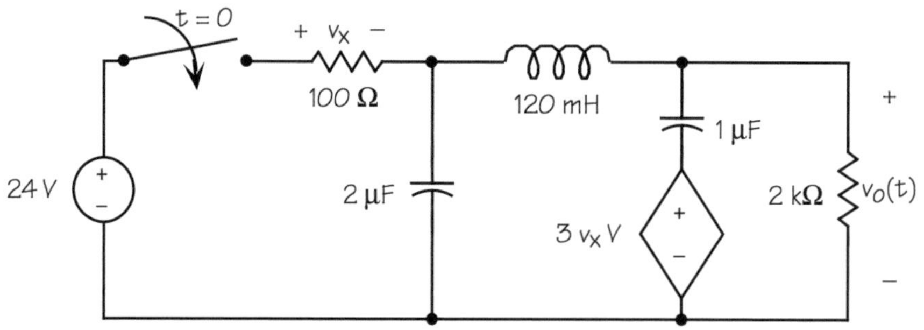

FIGURE 2.30: Examples XVI and XVII.

Note in the first two equations that I have written "0" after the integral to remind me that the initial current through the inductor is 0. These two equations apply Kirchhoff's current law at the nodes labeled v_1 and v_o.

The third and fourth equations bring in the constraint of the dependent source by defining v_x and hence v_2.

The last two equations are the initial values of the capacitor voltages. The initial value of the voltage across the 1-μF capacitor is given in terms of the difference between the voltages at its terminals.

The result isn't very pretty:

$$v_o(t) = 22.86 + 9.64e^{-2156t} + e^{-1672t}(39.50\cos 4183t - 73.92\sin 4183t)\,\text{V for } t > 0.$$

2.4.7   Example XVII

Redo the solution of the previous circuit (Fig. 2.30) using mesh analysis. I defined three mesh currents, i_1, i_2, and i_3, from left to right, all clockwise. The equations are

$$-24 + 100i_1 + \frac{1}{2\times 10^{-6}}\int_0^t (i_1 - i_2)dx + 0 = 0,$$

$$\frac{1}{2\times 10^{-6}}\int_0^t (i_2 - i_1)dx + 0 + 120\times 10^{-3}\frac{di_2}{dt} +$$

$$\frac{1}{1\times 10^{-6}}\int_0^t (i_2 - i_o)dx + 0 + 3v_x = 0,$$

$$-3v_x + \frac{1}{1\times 10^{-6}}\int_0^t (i_o - i_2)dx + 0 + 2\times 10^3 i_o = 0,$$

$$v_x = 100i_1,$$

$$i_2(0) = 0.$$

The first three equations are the mesh equations. Again, I have written "0" after each integral to remind me that the initial values of the capacitor voltages are 0.

One equation is needed to define v_x for the dependent source. There is one initial value for the derivative.

The result is the same nonpretty mess as before.

2.4.8   Example XVIII

The voltage v_o in the circuit of Fig. 2.31 is to be found in the sinusoidal steady state.

First, let's convert the circuit to the phasor domain. The impedances are

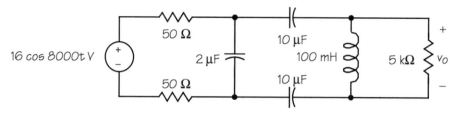

FIGURE 2.31: Examples XVIII and XIX.

$$\omega = 8000,$$
$$Z_{C2} = 1/(j \times 8000 \times 2 \times 10^{-6}) = -j62.5\,\Omega,$$
$$Z_{C10} = 1/(j \times 8000 \times 10 \times 10^{-6}) = -j12.5\,\Omega,$$
$$Z_L = j \times 8000 \times 100 \times 10^{-3} = j800\,\Omega.$$

I also converted the voltage source to a magnitude of 16 V with a phase angle of 0.

This circuit has five nodes. I have chosen the reference node at the bottom left, then labeled the rest of the nodes clockwise from the top left: +16, V_1, V_2, V_3, and V_4. The equations are straightforward:

$$\frac{V_1 - 16}{50} + \frac{V_1 - V_4}{Z_{C2}} + \frac{V_1 - V_2}{Z_{C10}} = 0,$$
$$\frac{V_2 - V_1}{Z_{C10}} + \frac{V_2 - V_3}{Z_L} + \frac{V_2 - V_3}{5 \times 10^3} = 0,$$
$$\frac{V_3 - V_4}{Z_{C10}} + \frac{V_3 - V_2}{Z_L} + \frac{V_3 - V_2}{5 \times 10^3} = 0,$$
$$\frac{V_4}{50} + \frac{V_4 - V_1}{Z_{C2}} + \frac{V_4 - V_3}{Z_{C10}} = 0,$$
$$V_o = V_2 - V_3.$$

The last equation relates the desired result, V_o, to the node voltages. The solution in the phasor domain is

$$V_o = 5.299 - j7.548 = 9.222\angle -54.9° \text{ V},$$

which translates back to the time domain as

$$v_o(t)_{sss} = 9.222\cos(8000t - 54.9°) \text{ V}.$$

2.4.9   Example XIX

Use mesh analysis to find the sinusoidal steady-state value of v_o in the same circuit (Fig. 2.31).

I have converted the circuit to the phasor domain as before. My mesh currents are I_1, I_2, and I_3 clockwise in the meshes from left to right.

The mesh equations are

$$-16 + 50I_1 + Z_{C2}(I_1 - I_2) + 50I_1 = 0,$$
$$Z_{C2}(I_2 - I_1) + Z_{C10}I_2 + Z_L(I_2 - I_3) + Z_{C10}I_2 = 0,$$
$$Z_L(I_3 - I_2) + 5 \times 10^3 I_3 = 0,$$
$$V_o = 5 \times 10^3 I_3.$$

The last equation gives the output V_o in terms of the mesh currents. The result is the same as before.

2.5   CIRCUIT DESIGN EXAMPLE

Design a simple circuit that will serve as an attenuator of signals presented to it. The circuit must meet four specifications:

- The circuit is to be simple and passive.
- The "gain" through the circuit will be a value between 0 and 1.
- The circuit is to match the source's internal resistance. (This means that the resistance "looking into" the circuit from the source terminals is R_s, measured with the source and R_s disconnected.)
- The circuit is to match the load resistance applied to it. (This means that the resistance "looking to the left into" the output terminals is R_L, measured with R_L disconnected.)

What will this circuit look like? Consider the specifications given. The first one tells us to make it entirely out of resistors (although we could perhaps consider inductors and capacitors). Op-amps are out.

The other three specifications are individual requirements. It would seem that each of these requirements would result in one algebraic equation. Since there are three such requirements, it seems reasonable to expect to need a minimum of three resistors. This gives us three "unknowns" to fulfill our three requirements.

Could the circuit require more than three resistors? Perhaps, but we'll leave the possible discovery of that until we have tried three.

How can we arrange three resistors in a circuit that has an input and an output? We will probably want a common ground path through the circuit, so only two arrangements come to mind, a T shape and a π shape. I'm going to choose the π shape as shown in Fig. 2.32, mainly because I want to. A T shape would probably work just as well.

Note in the circuit drawing that I have already set up the circuit for nodal

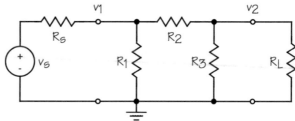

FIGURE 2.32:   Possible attenuator.

analysis. Also note that the source includes R_s, its internal resistance. The input voltage to the attenuator is v_1.

Now let's reduce the specifications to mathematical statements:

• The "gain" (call this A) is v_2 / v_1. I will compute this by solving two node equations:

$$\frac{v_1 - v_s}{R_s} + \frac{v_1 - v_2}{R_2} + \frac{v_1}{R_1} = 0,$$

$$\frac{v_2 - v_1}{R_2} + \frac{v_2}{R_3} + \frac{v_2}{R_L} = 0.$$

Solving these gives the result

$$A = \frac{v_2}{v_1} = \frac{R_3 R_L}{R_2 R_3 + R_2 R_L + R_3 R_L}.$$

• The input match requires the value of R_{in} (the resistance "seen" looking into the v_1 terminals from the left). This is most easily found by parallel and series resistances. R_L and R_3 are in parallel. These are in series with R_2. This group is in parallel with R_1. So the input resistance is

$$R_{in} = \frac{R_1 \left(R_2 + \dfrac{R_3 R_L}{R_3 + R_L} \right)}{R_1 + R_2 + \dfrac{R_3 R_L}{R_3 + R_L}}.$$

- Likewise, the output match requires the value of R_{out} (the resistance "seen" looking into the v_2 terminals from the right with v_s replaced by a short circuit). This also is found using parallel and series but from the right:

$$R_{out} = \frac{R_3\left(R_2 + \dfrac{R_1 R_s}{R_1 + R_s}\right)}{R_3 + R_2 + \dfrac{R_1 R_s}{R_1 + R_s}}.$$

These equations can be solved by setting $R_{in} = R_s$ and $R_{out} = R_L$. The result of doing this isn't too awful:

$$R_1 = \frac{R_s(A^2 R_s - R_L)}{A^2 R_s - 2A R_s + R_L},$$

$$R_2 = -0.5 \frac{A^2 R_s - R_L}{A},$$

$$R_3 = \frac{R_L(A^2 R_s - R_L)}{A^2 R_s - 2A R_L + R_L}.$$

So far, so good. Let's try this for some obvious case. Suppose the source is an audio amplifier with a line output of 600 Ω and the load is to be 600 Ω. Let's see what happens if I want an output 0.75 as large as the input: The numerical results are

$$R_1 = 4.2 \text{ k}\Omega, \quad R_2 = 175 \ \Omega, \quad R_3 = 4.2 \text{ k}\Omega.$$

Fig. 2.33 shows the circuit with all values.

Can I choose just any values? For example, suppose the amplifier is still 600 Ω but the load is an 8-Ω speaker and the gain is to be 0.75. Here are the results:

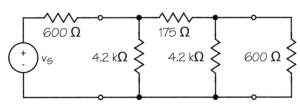

FIGURE 2.33:   Result for $R_S = R_L = 600 \ \Omega$, $A = 0.75$.

$$R_1 = 356.5 \ \Omega, \quad R_2 = -219.7 \ \Omega, \quad R_3 = -7.9 \ \Omega.$$

Oh oh, negative resistors! That can't be, so this is a circuit that cannot be built. What can be built? I experimented with this problem, using Maple. I found (by plotting and by trial-and-error) that if the gain is to be 0.75, R_L is constrained to the range

$$0.9375R_s < R_L < 1.125R_s$$

if we are not to have any negative resistances.

So is this design practical? Perhaps, but only for circuits for which the source resistance and the load are already about equal.

2.6   SUMMARY

The methods of nodal analysis and mesh analysis help us solve circuits without two equations for every branch in the network. The reduction is often large: eight branches would require 16 equations, but if there are only four nodes, the circuit needs only three equations.

Nodal analysis is important in computer algorithms. We generally think of circuits as having elements that are connected between nodes. So the concept of a branch is important. While both node and mesh equations are of about equal complexity, it seems to be rather hard to write procedures that can find loops. Hence the node method, which requires finding branch currents, is the preferred computer technique. The popular circuit simulation program SPICE uses primarily nodal analysis.

OK, so nodal analysis is popular with computers, but you don't think you are a computer? How should you choose a method? I can think of several ways to make this decision:

- I like nodal analysis because I learned it first.
- Mesh analysis is more fun.
- The test says to use nodal analysis.
- I'll choose the one with the smaller number of equations.

You might guess that I will suggest the last reason is the best. Well…not exactly. I often choose nodal analysis because I like it better when I type equations into Maple. Even when mesh might be an equation or two shorter. But the proper way to make the decision is to count nodes and count meshes and compare Nodes−1 :: Meshes.

You might have noticed some parallels between the two methods. They are, in fact, very parallel. While we won't go into it here, *duality* converts one form to the other. For every operation in nodal analysis that works on nodes, there is a dual operation in mesh analysis that works on meshes.

You can convert a problem to the dual problem by exchanging words: current and voltage, resistance and conductance, node and mesh (or loop), open and short. I'll end duality with a question. In mesh analysis I omitted Step 1. What is the dual of Step 1 in nodal analysis? Where is the reference mesh?

What's important in this chapter? If I could give just one answer, it would have to be nodal analysis. It works for all circuits (mesh analysis requires circuits to be planar). It is the method of choice for computer applications that start with the circuit itself.

So can you ignore mesh analysis? Nope! There are situations where it is better or neater or quicker or....

What's coming next? The formal solution of circuits is often overkill. We sometimes need only some insight into how a circuit behaves. At other times all we want to know is what the circuit does to an external load. So the techniques in the next chapter are going to give us some new approaches.

Keep in mind, though, that even with the nodal and mesh analysis that we have learned, plus the techniques in the next chapter, we still must be able to write with 100% reliability a mathematical description of a circuit. That was my goal in the first chapter, and it still it.

CHAPTER 3

Useful Theorems: Formal is Overkill?

The formal methods of mesh and nodal analysis that we have just learned will always work (with one restriction on meshes). These methods are very useful for complicated circuits. Nodal analysis lends itself very nicely to computer applications.

But methods can be overkill. We often deal with rather simple circuits that can be analyzed more simply. Sometimes all we need is a general feel for what is happening and don't need a complete analysis.

This chapter describes a number of simplification schemes that can make our circuit-analysis job easier. But they all depend on the circuit's being linear, so we'll start with that. Then we'll look at

- proportionality, which allows us to "guess" at results;
- superposition, which considers the effects of several sources individually;
- reciprocity, which you are already familiar with in a practical way;
- source transformations, which enable us to reduce a circuit to fewer components;
- Thévenin's and Norton's theorems, which formalize the process of source transformations; and
- maximum power transfer.

We start with linearity.

3.1 LINEARITY

Linear is everything! Our life would be more complicated if our circuits weren't linear. But the bad news is that circuits are not linear! Now what?

We assume that we are working with linear elements, even though we know that they aren't. But as long as we keep elements operating in the "right" ranges of current and voltage, we can have circuits whose response is very close to linear.

What does "linear" mean in a circuit? Suppose a circuit has an input v_{in} that produces a certain output v_{out}. If the circuit is linear, then an input of av_{in} will produce an output of av_{out}, no matter what value is assigned to a.

Linearity means even more. If an input v_{1in} produces an output v_{1out}, and an input v_{2in} produces an output v_{2out}, then an input of $av_{1in} + bv_{2in}$ will produce an output of $av_{1out} + bv_{2out}$.

Is linearity useful? Gosh, yes! It keeps the algebra simple when we write equations. It keeps the differential equations simple, too. And it also means that solutions usually exist for our circuits. (In nonlinear systems, solutions may not exist.)

3.1.1   Example I

Consider the circuit of Fig. 3.1. It is a linear circuit if all of its elements are linear. There is one voltage source, which we call ideal, that produces 30 V under all conditions. There are five resistors that we are going to say are linear, so the circuit is composed of linear elements, which makes it linear.

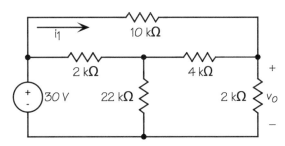

FIGURE 3.1:   Linear circuit.

I have analyzed this circuit and found that v_o = 10 V and that i_1 = 2 mA.

Suppose I want to know what these values are if I double the source to 60 V? Since the circuit is linear, I merely double the results: v_o = 20 V, i_1 = 4 mA.

What happens to the power delivered to the 2-kΩ resistor at the right? When the source is 30 V, the voltage across the resistor is 10 V. Hence the power delivered to that resistor is $10^2/2$ = 50 mW. For the 60-V source the voltage is 20 V, so the power delivered is $20^2/2$ = 200 mW. Power is *not* a linear function, so linearity does not apply!

Linearity gives me something more, though. Suppose I need to have an output voltage v_o = 15 V. What should the voltage source be for this? I know that 30 V in yields 10 V out. My desired output is half again as much, namely, 15 V. Linearity says make the input half again as much, or 45 V.

**Linearity is fundamental
to everything in this chapter!**

3.2   PROPORTIONALITY

Proportionality requires a linear circuit. It kind of reverses the direction we've been taking when analyzing circuits. It says, let's guess at a result somewhere in the circuit. Then we'll see if that value meets the other circuit constraints. If it doesn't, we'll use linearity to adjust our guess.

3.2.1 Example II

Consider the circuit we just saw in Fig. 3.1. We know that a source of 30 V will produce an output voltage of 10 V. Suppose I want to know what source to use for an output voltage of 15 V. I can get this through linear considerations:

$$15/10 = 1.5,$$
$$v_s = 30 \times 1.5 = 45 \text{ volts.}$$

Since the output is to be half again as large, so also must the input be half again as large.

The method of proportionality can be used to "solve" some circuits rather easily. It works best if the circuit fits three criteria:

- it looks like a ladder (actually, a ladder lying on its side),
- a single source is at one end, and
- the output is on the other end.

The basic idea is to guess at the answer, then work backward toward the source. When you get to the source, see how far off you are. Then adjust the "guessed" answer to get the correct source value.

3.2.2 Example III

In the circuit in Fig. 3.2, we want the output voltage v_o. While we can solve for this directly, let's use proportionality instead.

I am going to make a guess at the value of v_o. But since I don't really know what it should be, I'll choose a simple value, like 1. Note in Fig. 3.3 that I have assigned v_o and also calculated the current that must flow through the 3-kΩ resistor.

Since I now know the current on the right, I can also compute the voltage across the 2-kΩ resistor as in Fig. 3.3.

Now I know the total voltage on the right side of the circuit. Since (using KVL) the voltage across the 1250-Ω resistor must be the same as the sum of the voltages across the 2-kΩ and 3-kΩ resistors, I can get

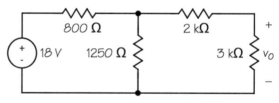

FIGURE 3.2: Circuit to analyze.

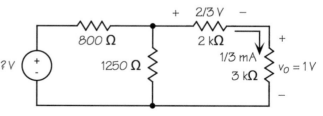

FIGURE 3.3: First guess and voltages.

that voltage. Then I can calculate the current through the 1250-Ω resistor. These are shown in Fig. 3.4.

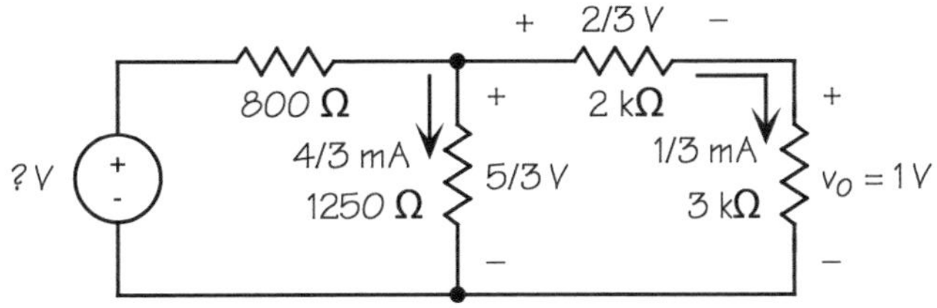

KCL will give me the current flowing from the left into the top node. Then I can calculate the voltage across the 800-Ω resistor. See Fig. 3.5.

But now KVL around the left mesh will give me the source voltage, which turns out to be 3 V as shown in Fig. 3.6.

But 3 V isn't the right source—it's supposed to be 18 V! No problem. 18 / 3 = 6, so I will multiply my original guess by 6 to get the correct value:

$$v_o = 6 \times 1 = 6 \text{ volts.}$$

The output doesn't really have to be at the far end of the circuit, either, as the next example will show.

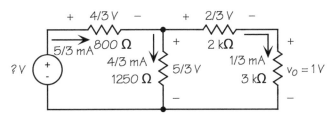

FIGURE 3.4: Middle voltage and current.

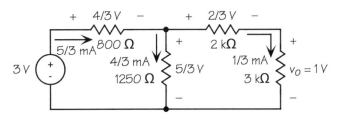

FIGURE 3.5: Left current and voltage.

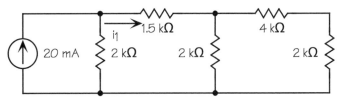

FIGURE 3.6: Resulting source voltage.

3.2.3   Example IV

In the circuit shown in Fig. 3.7, I would like to know the value of the current i_1.

While I could start by guessing what i_1 should be, it's easier to start at the far end of the circuit again. Fig. 3.8 shows what I have done, beginning with a guess for the voltage at the right-hand end.

FIGURE 3.7: Another ladder network.

This time I guessed the output voltage to be 2 V because that kept fractions out of the first calculations. The current on the right is 1 mA, which makes the voltage across the 4-kΩ resistor 4 V.

KVL gives 6 V across the middle 2-kΩ resistor, so the current down through it is 3 mA. A KCL computation at the top of the 2-kΩ resistor yields 4 mA through the 1.5-kΩ resistor.

Now I know the voltage across the 1.5-kΩ resistor (6 V), so I can use KVL to get the voltage across the left-hand 2-kΩ resistor (12 V).

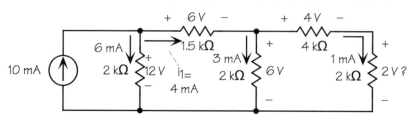

FIGURE 3.8: Guessed values.

Hence the current through that resistor is 6 mA.

Applying KCL once more, I find the source must be 10 mA to yield 2 V at the right and 4 mA for i_1. But the source is supposed to be 20 mA. So i_1 is

$$i_1 = (20/10) \times 4 = 8 \text{ mA}.$$

That's proportionality. It relies on linearity and gives us a way of getting results in some situations.

3.3   SUPERPOSITION

Superposition requires a linear circuit. It is fairly easily stated, but we'll do better when we see an example:

> **In a linear circuit,**
> **the voltage or current at any point**
> **is the sum of the voltages or currents**
> **that result from each source acting alone.**

How about that? A simple example should illustrate what this says.

3.3.1   Example V

The circuit in Fig. 3.9 has two sources. I want to see what each of them does to the output voltage v_o.

I'll start by turning off the 12-mA source and find out what v_o is when only the 20-V source is on. Hmmmm, but how do I "turn off" a source? Take a look at Fig. 3.10.

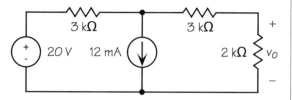

FIGURE 3.9: Circuit with two sources.

If a voltage source is to be "turned off," that would imply that it produces zero voltage. So it must insure that the voltage between two nodes is zero. That can be done with a short circuit.

Similarly, if a current source is to be "turned off," it must produce zero current. So it must insure that no current flows through its branch. That can be done with an open circuit.

FIGURE 3.10: Turning off sources.

Now, knowing all that, consider Fig. 3.11 where I have indicated the turned-off current source with an open circuit.

This is a voltage divider, so the result is

$$v_{o20} = 20\frac{2}{2+3+3} = 5 \text{ V}.$$

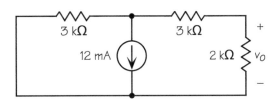

FIGURE 3.11: Circuit with current sources off.

So the voltage v_o due to the 20-V source acting alone is 5 V.

Now let's turn off the voltage source instead, as shown by the short circuit in Fig. 3.12.

Gosh, this is a current divider. If we define the current i_o as being downward through the 2-kΩ resistor, the result is

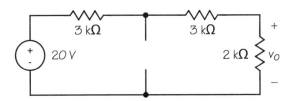

FIGURE 3.12: Circuit with voltage sources off.

$$i_{o12} = -12\frac{3}{3+3+2} = -4.5 \text{ mA},$$
$$v_{o12} = 2 \times i_{o12} = 2 \times (-4.5) = -9 \text{ volts}.$$

Superposition says we can find v_o by summing the v_o due to the 20-V source and the v_o due to the 12-mA source:

$$v_o = v_{o20} + v_{o12} = 5 - 9 = -4 \text{ volts}.$$

Ah, I know what you are thinking! Gee, you say, I could have found v_o by nodal analysis without all this superposition stuff. True. So what's this all good for?

Superposition can simplify the analysis of some circuits. But more important, it can allow us to "pick out" the effect of a particular source on some current or voltage. This gives us a way, for example, of determining how sensitive a certain voltage might be to changes in the output of one source, a source that might be a sensor of some kind.

Here's an example with a different kind of question.

3.3.2 Example VI

For the circuit shown in Fig. 3.13, find the value of the source v_c that will exactly reduce the output voltage v_o to zero.

The source v_c might be a source that will be used to control the output of the circuit. I'll use superposition to find the voltage v_o due to the 17-V source acting alone (Fig. 3.14).

Note that I have replaced the top source with a short circuit to guarantee that the voltage between its two nodes remains zero. Now combine the two top resistors (Fig. 3.15).

Since this is a ladder network with a source at one end and an output at the other, I used proportionality to find that $v_o = 6$ V due to the 17 V source acting alone.

This result tells us that the 17 V source causes v_o to be 6 V. Hence the control source v_c must be of such a voltage that it exactly cancels the 6 V provided by the 17-V source.

Another way to look at this is to say that v_c acting alone must make $v_o = -6$ V. That's what I am going to do to finish this problem. Fig. 3.16 is our circuit with the control source active. The 17 V source has been killed and replaced by a short circuit.

Let's combine the two resistors on the left, redraw the circuit, and label the nodes for nodal analysis (Fig. 3.17).

This circuit is a good one for the review of nodal analysis because it has a voltage source in a strange place. I've used v_o for one of the node voltages (with the reference node at the negative end of v_o). At the top is an assigned

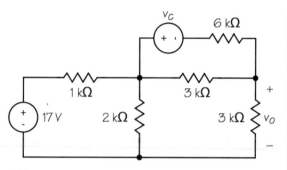

FIGURE 3.13: Circuit with "control source".

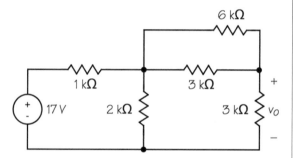

FIGURE 3.14: Circuit with control source dead.

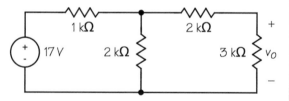

FIGURE 3.15: Simplified circuit.

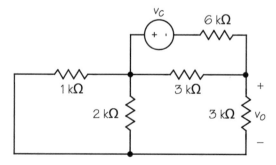

FIGURE 3.16: Circuit with 17-V source dead.

node voltage v_1. The node on the upper left is labeled $v_1 + v_c$ because it is determined by the node voltage v_1 and the source v_c.

The equation for the node v_o is an easy one to write. The other node is messier because of the source v_c. The current leaving v_1 to the right is easy, but we have to go around to the other side of the v_c source to get the currents leaving to the left.

Note the third equation, which sets v_o to the required -6 V:

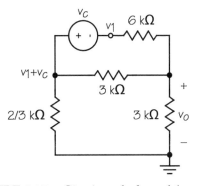

FIGURE 3.17: Circuit ready for nodal analysis.

$$\frac{v_o - v_1}{6} + \frac{v_o - (v_1 + v_c)}{3} + \frac{v_o}{3} = 0,$$

$$\frac{v_1 - v_o}{6} + \frac{(v_1 + v_c) - v_o}{3} + \frac{(v_1 + v_c)}{2/3} = 0,$$

$$v_o = -6 \text{ V}.$$

Solving these equations for v_c yields $v_c = 34$ V. So the control source must be at 34 V to exactly cancel the effect of the 17-V source.

3.3.3 Example VII

The circuit of Fig. 3.18 is excited by the source $v_i(t)$ that has the periodic waveform shown in Fig. 3.19. Find the sinusoidal steady state portion of the output $v_o(t)$.

The input signal can be described using terms of the Fourier series,[1] each of which is a sine wave at a different frequency. The period T of the signal is 1 ms, so the radian frequency ω_o is $2\pi/10^{-3} = 2\pi10^3$ rad/s. The first three terms of the Fourier series are

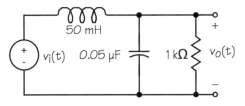

FIGURE 3.18: Example VII.

$$v_i(t) \approx 6.366 \sin\omega_o t + 2.122 \sin 3\omega_o(t) + 1.273 \sin 5\omega_o(t) \text{ V}.$$

The plot of Fig. 3.20 shows that this isn't too bad an approximation to the square wave.

---

[1] This is a topic beyond our course right now, so I'll do the Fourier work for you.

Now we have a problem. The excitation is composed of several frequencies. Yet what we have learned about impedances in the sinusoidal steady state says that we must have a single frequency. After all, inductors and capacitors have impedances that depend on frequency.

Superposition is our salvation. The circuit is linear. Therefore we can find the output due to each of the sinusoids acting independently. In effect, we are treating the sum of the three sinusoids as three different sources, all in series. After we have found the output due to each one, we can sum them (superimpose them) to get $v_o(t)$.

Fig. 3.21 shows the circuit transformed into impedances: the impedance of the inductor is $j\omega L$ and the impedance of the capacitor is $1/j\omega C$.

Now let's find V_o in terms of V_i by using a voltage divider with the resistor and the capacitor in parallel:

$$V_o = \frac{\dfrac{(-j10^6/0.05\omega) \times 1000}{(-j10^6/0.05\omega) + 1000}}{\dfrac{(-j10^6/0.05\omega) \times 1000}{(-j10^6/0.05\omega) + 1000} + j0.05\omega} V_i$$

$$= \frac{0.4 \times 10^9}{0.4 \times 10^9 - \omega^2 + j20 \times 10^3 \omega} V_i.$$

This is the expression for V_o for any input frequency. I'll evaluate this for each of the three frequencies represented in the input $v_i(t)$:

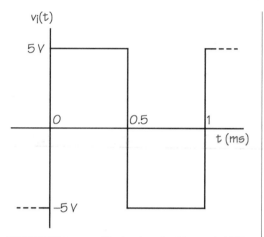

FIGURE 3.19: Excitation for Example VII.

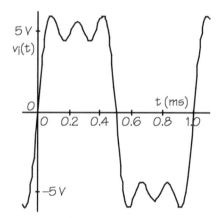

FIGURE 3.20: Fourier approximation for $v_i(t)$.

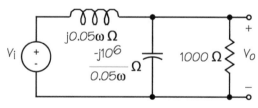

FIGURE 3.21: Example VII as impedances.

$\omega = 2\pi10^3$ rad/s	$V_{i1} = 6.366$ V	$V_{o1} = 6.670\,\underline{/-19.2°}$ V
$\omega = 6\pi10^3$ rad/s	$V_{i1} = 2.122$ V	$V_{o1} = 2.236\,\underline{/-83.2°}$ V
$\omega = 10\pi10^3$ rad/s	$V_{i1} = 1.273$ V	$V_{o1} = 0.592\,\underline{/-133.1°}$ V

Transforming these back to time functions and superimposing them, I get the result of

$$v_o(t) \approx 6.670\sin(2000\pi t - 19.2°)$$
$$+2.236\sin(6000\pi t - 83.2°) + 0.592\sin(10000\pi t - 133.1°)\,\text{V}.$$

What does this look like? The plot in Fig. 3.22 shows the result. Yes, this circuit has altered the waveform, but the important part of this example is that we can apply superposition in the sinusoidal steady state when we have multiple sources of different frequencies.

Superposition is a circuit-analysis technique that comes in handy now and then. The next topic, reciprocity, is likewise handy—now and then.

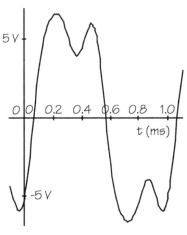

FIGURE 3.22: Output $v_0(t)$.

3.4 RECIPROCITY

Reciprocity requires a linear circuit. But even when the circuit is linear, it's a concept that is rather hard to get a feel for. Even after you know what it says, you are left to wonder where you'll see it used. Don't worry, it will show up one day!

Consider a circuit (linear!) with a source and an ammeter. Reciprocity says that exchanging the source and the ammeter won't change the reading on the meter. Let's see this in an example.

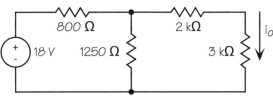

FIGURE 3.23: Circuit for Example III.

3.4.1 Example VIII

Here's the circuit of Example III again (Fig. 3.23). We found in Example III that the voltage across the 3-kΩ resistor was 6 V, so $i_o = 6/3 = 2$ mA.

Now let's exchange the positions of the 18-V source and the "ammeter." Fig. 3.24 shows this trade and the circuit is marked for mesh analysis.

Let's write the mesh equations. I've drawn the meshes counterclockwise this time and I'll apply KVL in that direction:

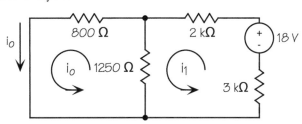

$$3000i_1 - 18 + 2000i_1 + 1250(i_1 - i_o) = 0,$$
$$1250(i_o - i_1) + 800i_o = 0.$$

FIGURE 3.24: Circuit with exchange completed.

Solving these yields i_0 = 2 mA, just what we got with things the other way around.

Great! So what? Do you mean you've never been in the position of needing to exchange the source and the output? Well, any ham radio operator knows that the antenna can be used for both transmitting and receiving (Fig. 3.25). And if you've messed around with an intercom system, you know that generally the same speaker that is used to produce the sound is also used for the microphone.

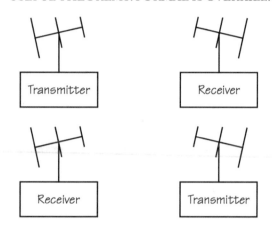

FIGURE 3.25: Reciprocal transmitter and receiver.

In the next section we'll take up a circuit element that displays reciprocity.

3.4.2 Mutual Inductance

When two inductors are placed in an arrangement where some of the magnetic flux of one coil passes through the other coil, we have coupled coils. These we represent in circuits as *mutual inductance*. Fig. 3.26 shows coupled coils with the labels that we need for circuits use.

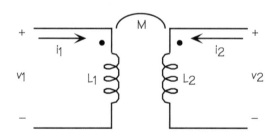

FIGURE 3.26: Coupled coils = mutual inductance.

Each of the coils has a *self-inductance*, here labeled L_1 and L_2. These are the inductances that the coils exhibit when they are completely separated.

The pair of coils also has a *mutual inductance*, here labeled M. The mutual inductance tells us the effect of one coil on the other. Note especially that M is the same in both directions, so this element is going to turn out to exhibit reciprocity.

What are those two dots? They tell us how to decide on the sign of the coupling effect. Here's a simple way to understand the dots:

> **A current flowing into a dot on one side**
> **will generate a voltage on the other side**
> **that is positive at the dot.**

Using the dots to establish the signs, I can write the equations for these coils. They look much like the relationship you have already learned for an inductor:

$$v_1 = L_1 \frac{di_1}{dt} + M \frac{di_2}{dt},$$

$$v_2 = M \frac{di_1}{dt} + L_2 \frac{di_2}{dt}.$$

Note the familiar $L \, di / dt$ terms. The two self-inductances are indeed inductors. Both mutual terms are positive here. For example, when i_2 is flowing into the dot at the top on the right, the voltage on the left has its + sign at the dot. Hence it adds to v_1.

What happens when we reverse one dot as shown in Fig. 3.27?

Consider, for example, that i_2 is now *not* flowing into its dot, so the voltage on the other side is *not* positive at the dot. Hence it will subtract from v_1. Here are the equations:

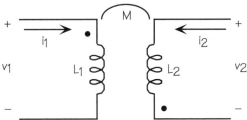

FIGURE 3.27:   Coupled coils with one dot reversed.

$$v_1 = L_1 \frac{di_1}{dt} - M \frac{di_2}{dt},$$

$$v_2 = -M \frac{di_1}{dt} + L_2 \frac{di_2}{dt}.$$

Both M terms are negative because the effect of the single reversal is two way: it affects both sides of the inductor in the same way.

The mutual inductance M is also an inductance like any that you have already studied. You'll learn more about this in electromagnetic fields. However, it is worth knowing that there is a coupling factor k that tells us how tightly the coils are coupled. This factor can't be less than zero (zero means totally uncoupled) and it can't be greater than one (one means perfectly coupled). Here's the relationship of k to the inductances:

$$M = k\sqrt{L_1 L_2}, \text{ where } 0 \le k \le 1.$$

3.4.3   Example IX

Let's use coupled coils in a circuit in the sinusoidal steady state. Fig. 3.28 shows such a circuit.

The first step is to convert everything to impedances. The mutual inductance (250 mH) converts to impedance just as does any other inductance. The result is shown in Fig. 3.29. I've marked the circuit for mesh analysis; mutual inductance lends itself better to mesh than nodal analysis. Don't let the single common line through the bottom of the circuit fool you! No current can flow on that single wire from the left half to the right half of the circuit because it has no way to return except over the same path.

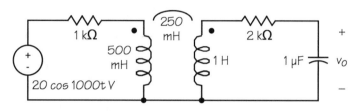

FIGURE 3.28: Sinusoidal steady state with mutual inductance.

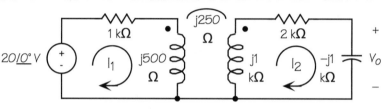

FIGURE 3.29: Circuit transformed.

Now write two mesh equations and solve them:

$$-20 + 1000I_1 + j500I_1 - j250I_2 = 0,$$
$$j1000I_2 - j250I_1 + 2000I_2 - j1000I_2 = 0,$$
$$V_o = -j1000I_2.$$

The self-inductance terms are handled in the same way that you've handled inductances before. It's the mutual terms that are tricky. In the first equation, as we go around clockwise, we encounter the self-inductance voltage, which is $j500I_1$. But there is a voltage across the $j500$ inductor caused by the current I_2. Since I_2 is *leaving* its dot on the right, it creates a voltage on the left that is *minus* at that dot. Hence this voltage shows up in the first equation as $-j250I_2$.

The same thing happens in the second equation. I_1 is entering its dot, so it creates a voltage on the right side that is plus at the dot. But our KVL path is going *up* through the inductor and hence the mutual voltage is negative.

Solving the equations gives

$$V_o = 2.181 \angle -25.87° \text{ volts.}$$

So mutual inductance (coupled coils) is just another circuit element. But it adds an extra term in each equation that involves the element.

3.5 SOURCE TRANSFORMATIONS

Source transformation requires a linear circuit. Our goal here is to show that we can trade a voltage source and series resistance for a current source and parallel resistance. And vice versa. This trade is useful in simplifying circuits and in understanding how some of them work.

3.5.1 Equivalence

Consider the two circuits shown in Fig. 3.30.

Under what conditions will these sources be equivalent? And what does *equivalent* mean?

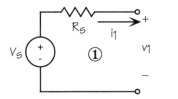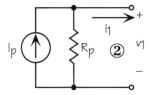

FIGURE 3.30: Two source arrangements.

Let's consider the term *equivalent* first. These two circuits will be considered equivalent if they "look" the same at the terminals. To say this another way, they will be equivalent if there is no test that you can perform *at the terminals* that will detect which source is behind those terminals.

Note that this testing excludes any but electrical tests at the terminals. You can't weigh the box or feel whether it is hot or look behind the terminals.

Remember that one condition is the circuits be linear. So if these are linear circuits, a plot of current versus voltage at the terminals must be a straight line. If such a plot is a straight line, then we need only two points to establish that line.

I'm going to choose two simple points: the open-circuit voltage and the short-circuit current. The graph of Fig. 3.31 shows the plots for my two circuits.

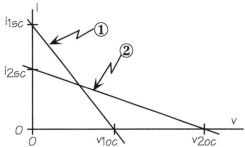

FIGURE 3.31: Terminal characteristics.

If I want these two sources to be equivalent, I must make the two lines lie right on top of one another. Hence the two circuits must have the same open-circuit voltage, and they must have the same short-circuit current:

$$v_{1oc} = V_s, v_{2oc} = I_p R_p, \text{ and make } v_{1oc} = v_{2oc};$$

$$i_{1sc} = \frac{V_s}{R_s}, i_{2sc} = I_p, \text{ and make } i_{1sc} = i_{2sc}.$$

These will all be true when

$$R_p = R_s,$$

$$I_p = \frac{V_s}{R_s}.$$

To say this another way, they both must have the same resistance and the V_s:I_p relationship must obey a kind of Ohm's law.

The two sources in Fig. 3.32 are equivalent:

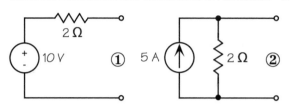

$$v_{1oc} = 10 \text{ V}, v_{2oc} = 5 \times 2 = 10 \text{ V};$$
$$i_{1sc} = 10/2 = 5 \text{ A}, i_{2sc} = 5 \text{ A}.$$

FIGURE 3.32: Equivalent sources.

I could use either one of them in a circuit and the *circuit* would never know the difference.

There is no test you can perform *at the terminals* that will determine which of these is connected into a circuit.

3.5.2 Example X

Let's use source trans-formations to simplify a circuit and find a result. In the circuit of Fig. 3.33 we are to find v_o.

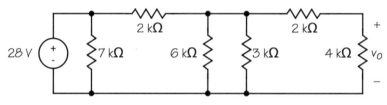

FIGURE 3.33: Circuit for source transformations.

While I'm going to use source transformations, I'll not just blindly attack the circuit! There are two simplifications to be made. Note the 7-kΩ resistor. It is directly across the 28-V source and therefore has no effect whatsoever on the rest of the circuit. It's just sitting there getting hot, because the voltage across it will always be 28. So I'll elim-inate it as I work to find v_o. I'll also combine the two parallel resistors into one. See Fig. 3.34.

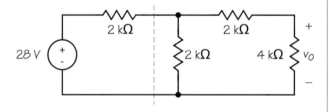

FIGURE 3.34: Circuit somewhat reduced.

Consider the 28-Vsource and the 2-kΩ resistor to the left of the dotted line. I'm going to use a source transformation to make them into a current source (28 / 2 = 14 mA) in parallel with a 2-kΩ resistor. The result is shown in Fig. 3.35.

Now I'll combine the two 2-kΩ resis-tors into a single 1-kΩ resistor and trans-

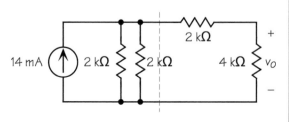

FIGURE 3.35: Circuit after one transformation.

form everything to the left of the dotted line. I'll get a voltage source (14 x 1 = 14 V) in series with a 1-kΩ resistor (Fig. 3.36).

Now we finish the job very easily by using a voltage divider, giving

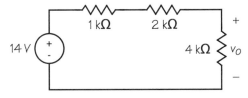

FIGURE 3.36:   Circuit after two transformation.

$$v_o = 14\frac{4}{1+2+4} = 8 \text{ volts}.$$

Source transformations can be useful in more complex cases, as we'll see in the next section.

3.6   THÉVENIN AND NORTON

Thévenin's theorem and Norton's theorem both require linear circuits. While there are formal, rather mathematical ways to introduce these, I'm choosing to do it in a practical way.

3.6.1   Example XI

Consider again the circuit that we used in the previous example, reproduced here as Fig. 3.37.

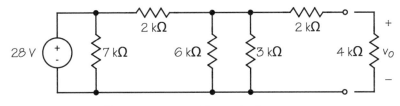

FIGURE 3.37:   Previous circuit with terminals.

I have sort of removed the 4-kΩ resistor as if it were a load. I want to study only the circuit to the left of the terminals. In fact, what I really want to do is reduce the whole circuit to a simple voltage-source-and-series-resistor connected to the terminals. I'll do this in four steps.

Step 1. Remove the 7-kΩ resistor as before and combine the 6-kΩ and 3-kΩ resistors (Fig. 3.38).

Step 2. Transform the 28-V source and 2-kΩ resistor (Fig. 3.39).

Step 3. Combine the two 2-kΩ resistors and transform the 14-mA source and 1-kΩ resistor that results (Fig. 3.40).

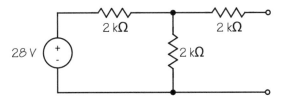

FIGURE 3.38:   Slightly simplified.

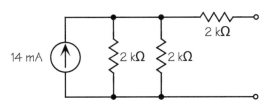

FIGURE 3.39:   First transformation.

Step 4. Combine the 1- and 2-kΩ resistors (Fig. 3.41).

I have now reduced the circuit to a very simple one, one which is equivalent to the original circuit *at the terminals*. Now I'll put the 4-kΩ load resistor back and compute v_o using a simple voltage divider (Fig. 3.42).

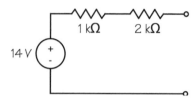

FIGURE 3.40:   Second transformation.

$$v_o = 14\frac{4}{3+4} = 8 \text{ volts.}$$

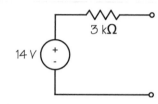

FIGURE 3.41:   Final result.

I hope it's no surprise that the result is the same!

Here's some news, though: The new circuit to the left of the terminals is the Thévenin equivalent of the original circuit. See, you didn't do anything different! The two circuits of Fig. 3.43 act exactly the same way at the terminals.

But we can get the Thévenin equivalent directly without all those transformations.

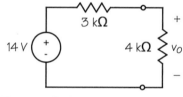

FIGURE 3.42:   Reduced circuit with original load.

3.6.2   Thévenin and Norton Directly

The Thévenin equivalent has two parameters, the voltage of the voltage source and the value of the series resistance. Let's start with the same example and do three things:

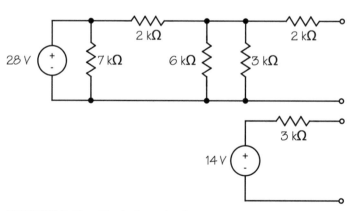

FIGURE 3.43:   Equivalent circuits.

- find the open-circuit voltage at the terminals,
- find the short-circuit current at the terminals, and
- find the resistance "looking into" the circuit at the terminals with the source dead.[2]

---

[2] I'll often say "dead" instead of "turned off."

Fig. 3.44 shows our circuit with preparations for finding the open-circuit voltage v_{oc}.

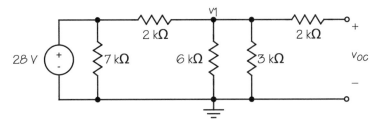

FIGURE 3.44: Circuit prepared for v_{oc}.

I wrote a single node equation and noted that, since there is no current through the 2-kΩ resistor on the right, v_1 and v_{oc} are the same voltage:

$$\frac{v_1 - 28}{2} + \frac{v_1}{6} + \frac{v_1}{3} = 0,$$

$$v_{oc} = v_1.$$

Solving, I get $v_{oc} = 14$ V.

Now here's the circuit again (Fig. 3.45), this time set up to find the short-circuit current.

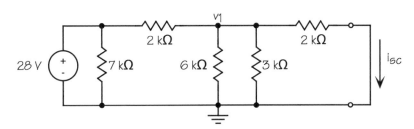

FIGURE 3.45: Circuit prepared for i_{sc}.

Again, I wrote a single node equation. The short-circuit current i_{sc} is through the 2-kΩ resistor on the right:

$$\frac{v_1 - 28}{2} + \frac{v_1}{6} + \frac{v_1}{3} + \frac{v_1}{2} = 0,$$

$$i_{sc} = \frac{v_1}{2}.$$

I get $i_{sc} = 14 / 3$ mA = 4.667 mA.

We can find the resistance by "looking into" the terminals with all the sources dead. This works only if there are *no de*pendent sources, because there is no way to kill those. Fig. 3.46 reminds us of what it means to kill a source.

Fig. 3.47 is our circuit with dead sources.

I have shown an arrow with a resistance label to indicate "looking into" the terminals. Note that the 7-kΩ resistor is gone because the short circuit of the dead voltage source makes it impossible to have any voltage across that resistance. Hence there can't be a current through it.

To solve for the resistance, I note that there are three resistors in parallel: 2 kΩ on the left, 6 kΩ, and 3 kΩ. These combine into a single 1-kΩ resistor. Add this to the 2-kΩ series resistor on the right and the result is $R_{Th} = 3$ kΩ.

The Thévenin equivalent is merely the open-circuit voltage applied to a voltage source in series with the value of R_{Th} found in the "looking in" step. Fig. 3.48 shows the result, which is the same as the one we got previously.

The Norton equivalent is the source transformation of the Thévenin equivalent. It is also the short-circuit current applied to a current source in parallel with R_{Th} already found. Fig. 3.49 shows the Norton equivalent.

It is important to note that

$$v_{oc} = R_{Th} i_{sc},$$

$$14 = 3 \times \frac{14}{3}.$$

This is an important result! It says that we can find the Thévenin equivalent (or the Norton equivalent) by finding *any two* of v_{oc}, i_{sc}, and R_{Th}, then computing the third one if needed.

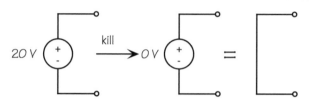
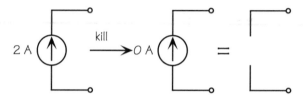

FIGURE 3.46: Killing sources.

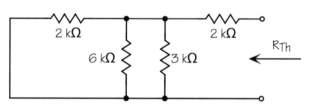

FIGURE 3.47: Circuit with dead sources.

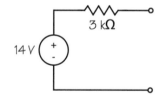

FIGURE 3.48: Thévenin equivalent.

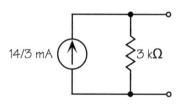

FIGURE 3.49: Norton equivalent.

3.6.3 Thévenin and Norton More Formally

The drawing in Fig. 3.50 reflects the general aspects of Thévenin's theorem.

Linear circuit A on the left is any circuit. While it must be linear, it can have anything in it: resistors, capacitors, inductors, mutual inductors, independent current and voltage sources, dependent current and voltage sources. It might have no sources. It might have no elements at all.

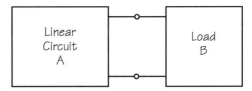

Load B on the right is anything. It doesn't even have to be linear.

FIGURE 3.50:   Thévenin in general.

Thévenin's theorem tells us how to make an equivalent of everything to the left of the terminals:

**As far as anything the Load B can determine,
Linear circuit A can be replaced by a voltage source in series with a resistance,
where the voltage source is the voltage measured at the terminals with Load B removed,
and where the resistance R_{Th} is the ratio of that voltage to the current
measured when the terminals are short circuited.**

The statement of Norton's theorem is almost the same, except that it yields a current source, whose value is the short-circuit current, in parallel with the same resistance.

Let's finish all this with four examples.

3.6.4 Example XII

In Fig. 3.51, find the Thévenin and Norton equivalents at the terminals a–b.

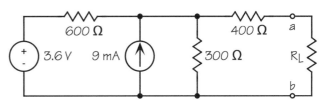

FIGURE 3.51:   Circuit for Example XII.

I'm going to mentally remove the load R_{L} and then find the voltage v_{ab} by superposition.

Start by killing the current source, which leaves an open circuit in its place. The 400-Ω resistor has no current through it and hence no voltage across it when the terminals are open circuited. The terminal voltage is found using a voltage divider:

$$v_{ab3.6} = 3.6\frac{300}{300+600} = 1.2 \text{ volts.}$$

Now put the current source back and kill the voltage source by replacing it with a short circuit. This puts the 600- and 300-Ω resistors in parallel. Again, the 400-Ω resistor has no current through it:

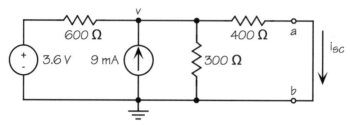

FIGURE 3.52: Circuit for short-circuit current.

$$v_{ab9} = 9 \times (0.6 \| 0.3) = 9 \times 0.2 = 1.8 \text{ volts.}$$

So the open-circuit voltage is the sum of the voltages of each source acting alone:

$$v_{oc} = v_{ab3.6} + v_{ab9} = 1.2 + 1.8 = 3.0 \text{ volts.}$$

Next, find the short-circuit current. Fig. 3.52 shows the circuit set up to find this. I'm going to write one node equation:

$$\frac{v - 3.6}{0.6} - 9 + \frac{v}{0.3} + \frac{v}{0.4} = 0,$$

$$i_{sc} = \frac{v}{0.4}.$$

The result is i_{sc} = 5 mA.

I'll find R_{Th} by calculation:

$$R_{Th} = \frac{v_{oc}}{i_{sc}} = \frac{3}{5} = 0.6 \text{ k}\Omega = 600 \ \Omega.$$

Fig. 3.53 shows both equivalent circuits. Either of these could replace the entire circuit to the left of the terminals a-b and the load resistor R_L would never know the difference.

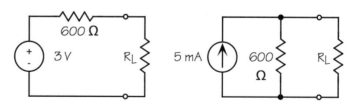

FIGURE 3.53: Thévenin and Norton equivalents.

Be sure you don't forget, though, that it is illegal, immoral, and crass to ask any questions about the circuit to the left and try to answer them by looking at the equivalents. For example, in the original circuit with the load disconnected, the resistors on the left are dissipating 30.6 mW. In the Thévenin equivalent with the load disconnected, the internal resistance dissipates no power, while in the Norton equivalent under the same conditions, the internal resistance dissipates 15 mW.

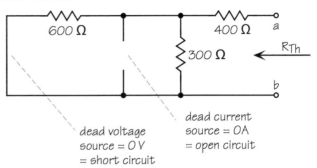

FIGURE 3.54: Circuit prepared for "looking in".

We could also have found R_{Th} by the "looking in" process because the circuit contains no dependent sources. Fig. 3.54 shows the circuit with its sources killed:

$$R_{Th} = 600\|300 + 400 = 600 \ \Omega.$$

3.6.5 Example XIII

Here's one more example, this time with a circuit (Fig. 3.55) that contains a dependent source.

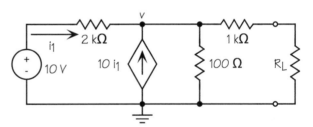

FIGURE 3.55: Circuit for Example XIII.

I'll find both v_{oc} and i_{sc} via nodal analysis because the circuit is messier than I want to consider otherwise. Here are the node equations for the open-circuit voltage. Keep in mind that there is nothing connected to the terminals and there is no current through the 1-kΩ resistor:

$$\frac{v-10}{2} - 10i_1 + \frac{v}{0.1} = 0,$$

$$i_1 = \frac{10-v}{2},$$

solution $\qquad v = v_{oc} = 3.548 \text{ V}.$

For the short-circuit current, remember that the terminals are shorted and I am looking for the current flowing through that short circuit:

$$\frac{v-10}{2} - 10i_1 + \frac{v}{0.1} + \frac{v}{1} = 0,$$

$$i_1 = \frac{10-v}{2},$$

solving, $\quad v = 3.333$ V,

so $\quad i_{sc} = \frac{v}{1} = 3.333$ mA.

I can't find R_{Th} by "looking in" so I'll use the ratio of the open-circuit voltage to the short-circuit current:

$$R_{Th} = \frac{v_{oc}}{i_{sc}} = \frac{3.548}{3.333} = 1.064 \text{ k}\Omega.$$

The equivalent circuit is shown in Fig. 3.56.

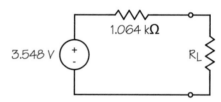

FIGURE 3.56: Equivalent circuit for Example XIII.

3.6.6 Example XIV

Thévenin and Norton can also do their work in the sinusoidal steady state. In the circuit shown in Fig. 3.57, we want to replace the circuit to the left of the terminals with its Thévenin equivalent.

We first convert the circuit to its representation in the world of phasors:

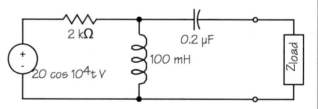

FIGURE 3.57: Example XIV for Thévenin.

$V_s = 20$ V,

$Z_L = j0.1 \times 10^4 = j1000 \, \Omega,$

$Z_C = \dfrac{-j}{0.2 \times 10^{-6} \times 10^4} = -j500 \, \Omega.$

Now let's find the open-circuit voltage by using a voltage divider:

$$V_{oc} = 20 \frac{j1000}{j1000 + 2000} = 8.94 \angle 63.4° \text{ V}.$$

The Thévenin impedance is gotten by series–parallel:

$$Z_{Th} = -j500 + \frac{j1000 \times 2000}{j1000 + 2000}$$

$$= -j500 + 400 + j800 = 400 + j300$$

$$= 500\angle 36.9° \ \Omega.$$

So the result of all this is the circuit shown in Fig. 3.58. The time-domain model of this circuit is shown in Fig. 3.59. Either one replaces the original circuit as far as the load Z_L is concerned.

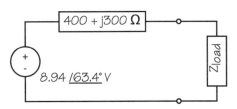

FIGURE 3.58: Thévenin equivalent for Example XIV.

3.6.7 Example XV

At the risk of going through too many examples, I still need to look at one more case. The circuit in Fig. 3.60 has no independent sources, so there is nothing to kill. (This is a very "academic" problem just to demonstrate a technique.)

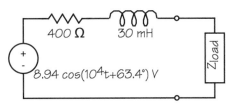

FIGURE 3.59: Thévenin equivalent in time domain.

We can't kill a dependent source. This makes the open-circuit voltage easy to find. If there is nothing connected to the terminals, the voltage v_x is zero. Hence the source is also zero. So the open-circuit voltage is also zero.

That was easy, but now what? Well, we could try for the short-circuit current. By the same argument, that's also zero! So the Thévenin resistance is 0/0, which can have a value.

The way we find that value is to apply energy to the circuit using a "test source." In Fig. 3.61 I have connected a current source of 1 mA to the circuit's terminals. (I chose 1 mA

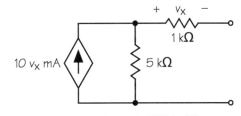

FIGURE 3.60: Example XV for Thévenin.

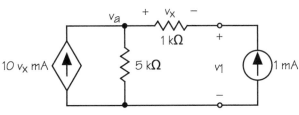

FIGURE 3.61: Example XV with test source.

because it's a simple number.) Then I'll calculate the voltage v_1 across the terminals. The Thévenin resistance will be that voltage v_1 divided by 1 mA.

By inspection,

$$v_x = -1 \times 1 = -1\,\text{V},$$
$$10v_x = -10\,\text{mA}.$$

Summing the currents leaving the top of the 5-kΩ resistor:

$$-(-10) + \frac{v_a}{5} - 1 = 0,$$
$$v_a = -45\,\text{V}.$$

Finally, writing KVL around the right-hand loop:

$$-v_a + v_x + v_1 = 0,$$
$$-(-45) + (-1) + v_1 = 0,$$
$$v_1 = -44\,\text{V}.$$

So I now can get the Thévenin equivalent resistance:

$$R_{Tb} = \frac{v_1}{1} = \frac{-44}{1} = -44\,\text{k}\Omega.$$

It's negative! Dependent sources can do that to you, but only if they have an external supply of energy. The Thévenin equivalent circuit is shown in Fig. 3.62.

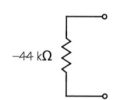

−44 kΩ

FIGURE 3.62: Thévenin equivalent of Example XV.

3.7 MATCHING

Let's go back to Example XII, shown in Fig. 3.63.

We found the Thévenin equivalent to be the circuit shown in Fig. 3.64.

How much power is delivered to the load resistor R_L? In fact, I'd like to know this for various values of R_L. Let's write an equation for P_L, the power delivered to R_L. I'll find the current in the loop, then square it and multiply it by R_L:

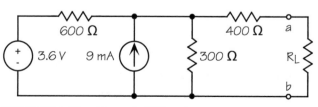

FIGURE 3.63: Example XII again.

$$i_L = \frac{3}{0.6 + R_L} \text{ mA},$$

$$P_L = i_L^2 R_L = \left(\frac{3}{0.6 + R_L}\right)^2 R_L \text{ mW}.$$

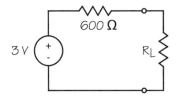

FIGURE 3.64: Thévenin equivalent.

It is interesting to plot this. The graph of Fig. 3.65 is P_L in milliwatts versus R_L in kilohms.

Interesting! There is a peak, which makes sense if you think of what the end values are:

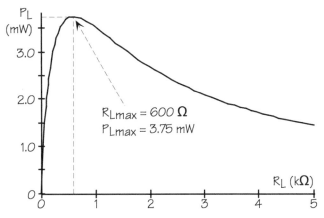

FIGURE 3.65: Power delivered to R_L.

- When $R_L = 0$, it absorbs no power, so the curve must reach the origin.
- When $R_L = \infty$, it absorbs no power because there is no current through it, so the curve must be asymptotic to the horizontal axis.
- Hence there must be a maximum (or minimum) somewhere in between.

The circuit for the general case is shown in Fig. 3.66.

The power delivered to R_L is easily calculated:

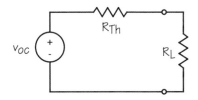

FIGURE 3.66: General Thévenin equivalent.

$$i_L = \frac{v_{oc}}{R_{Th} + R_L},$$

$$P_L = i_L^2 R_L = \left(\frac{v_{oc}}{R_{Th} + R_L}\right)^2 R_L.$$

It is left as an exercise for the student (arghhh!) to maximize this function with respect to R_L. The result will be that $R_L = R_{Th}$.

We can see this on the graph of Fig. 3.65, where the peak looks like it occurs at 600 Ω, the Thévenin equivalent resistance.

This is called *matching*, where maximum power is transferred from the source to the load. This applies only when the source itself is *fixed*. In other words, if you can change the source, get out of it all the losses you can. But if it is fixed, you'll get the best results into the load by matching $R_L = R_{Th}$.

Matching can also be done with imped-
ances in the sinusoidal steady state. In Fig. 3.67 I
have repeated the Thévenin equivalent of Example
XIV. Let's find the value of Z_{load} that will extract
maximum power from this circuit.

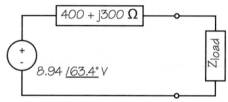

FIGURE 3.67:   Example XIV repeated.

While the obvious solution might be to say
that Z_{load} should equal the Thévenin equivalent
impedance Z_{Th}, we'd come up short of power. Part
of Z_{Th} is imaginary, and imaginary terms can be positive or negative. So why not choose the
imaginary part of Z_{load} to be negative so that it cancels the positive imaginary part of Z_{Th}?
That would reduce the total series impedance in the circuit to a minimum and maximize the
current.

So for a matched load for Example XIV,

$$Z_{load} = 400 - j300 \, \Omega$$

and hence the load voltage is just half the source voltage:

$$V_{out} = 0.5 \times 8.94 \angle 63.4° = 4.47 \angle 63.4° \text{ V}.$$

The resultant load is shown in Fig. 3.68.

3.8   SOME MORE EXAMPLES

The following examples cover proportionality, superposition, coupling, and
Thévenin.

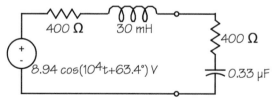

FIGURE 3.68:   Example XIV with matched load.

3.8.1   Example XVI

Redo Example XVIII from Chapter 2 using proportionality. (The circuit is redrawn in Fig.
3.69 with labels on voltages and currents.)

Conversion of the circuit to the phasor domain yields the following impedances as
before:

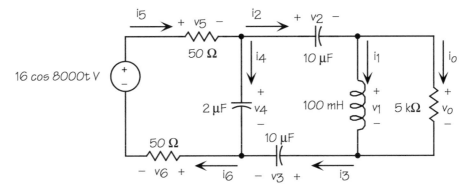

FIGURE 3.69:   Example XVI; Proportionality.

$$\omega = 8000,$$
$$Z_{C2} = 1/(j \times 8000 \times 2 \times 10^{-6}) = -j62.5\,\Omega,$$
$$Z_{C10} = 1/(j \times 8000 \times 10 \times 10^{-6}) = -j12.5\,\Omega,$$
$$Z_L = j \times 8000 \times 100 \times 10^{-3} = j800\,\Omega.$$

I will start by assuming that the output v_o is 1 V. The individual steps are laid out in the following set of equations:

$$v_{o-try} = 1\,\text{V},$$
$$i_o = v_{o-try}/5 \times 10^3 = 0.2\,\text{mA},$$
$$v_1 = v_{o-try} = 1\,\text{V},$$
$$i_1 = v_1/Z_L = -j1.25\,\text{mA},$$
$$i_2 = i_1 + i_o = 0.2 - j1.25\,\text{mA},$$
$$i_3 = i_2 = 0.2 - j1.25\,\text{mA},$$
$$v_2 = Z_{C10}i_2 = -15.62 - j2.50\,\text{mV},$$
$$v_3 = v_2 = -15.62 - j2.50\,\text{mV},$$
$$v_4 = v_3 + v_1 + v_2 = 0.9688 - j0.0050\,\text{V},$$
$$i_4 = v_4/Z_{C2} = 0.08 + j15.50\,\text{mA},$$
$$i_5 = i_2 + i_4 = 0.28 + j14.25\,\text{mA},$$
$$i_6 = i_5 = 0.28 + j14.25\,\text{mA},$$
$$v_5 = 50i_5 = 0.014 + j0.713\,\text{V},$$
$$v_6 = 50i_6 = 0.014 + j0.713\,\text{V},$$
$$v_{s-try} = v_6 + v_4 + v_5 = 0.9968 + j1.4200\,\text{V}$$
$$= 1.735\angle 54.9°\,\text{V}.$$

The result of these calculations is that the source should be 1.735$\underline{/54.9°}$ V if the output v_o is going to be 1. Of course, the source voltage is not, instead it is 16$\underline{/0°}$ V. Hence the actual output voltage is

$$v_{o-actual} = v_{o-try} \frac{16}{1.735} = 9.222 \text{ V},$$

with a phase angle of

$$v_{o-angle} = -54.9°.$$

The sinusoidal steady state output voltage in the time domain is

$$v_o(t) = 9.222 \cos(8000t - 54.9°) \text{ V}.$$

Is that an easier way to work this problem, easier than using mesh or nodal analysis, easier than firing the big gun? It depends on what one means by "easier"! For example, the steps of the proportionality are well suited to a simple calculator.

3.8.2 Example XVII

A certain signal can be represented as a sum of the terms of a Fourier series. This signal must be filtered to suppress to some extent the fundamental and the second harmonic without too much attenuation of the third and higher harmonics.

$$v_{signal} = 5\cos(5,000t) + 0.2\cos(10,000t + 40°) +$$
$$3\cos(15,000t + 80°) + 0.08\cos(20,000t + 120°) +$$
$$1.2\cos(25,000t + 160°) \text{ V}.$$

The engineer who is working on the project has designed the filter circuit of Fig. 3.70. Our job is to see how well this meets these rather vague specifications.

First, though, let's be sure that we know what the harmonics are. Here, the fundamental frequency (also called the first harmonic) is 5000 rad/s. The second harmonic has a frequency of 2 × 5000 = 10,000 rad/s; the third is 3 × 5000 = 15,000 rad/s.

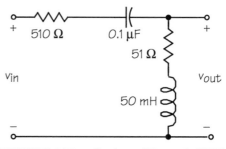

FIGURE 3.70: Design of Example XVII.

Note that the second harmonic is small compared with its neighbors, so I will ignore it in my analysis. What I want to get is the attenuation of the third harmonic relative to the first and compare it with the ratio of the third to the first before attenuation.

I need the circuit converted to the phasor domain for two frequencies, those of the first and third harmonics:

$$\omega = 5000,$$
$$Z_{C1} = 1/(j \times 5000 \times 0.1 \times 10^{-6}) = -j2000 \, \Omega,$$
$$Z_{L1} = j \times 5000 \times 50 \times 10^{-3} = j250 \, \Omega.$$
$$Z_{C3} = 1/(j \times 3 \times 5000 \times 0.1 \times 10^{-6}) = -j666.7 \, \Omega,$$
$$Z_{L3} = j \times 3 \times 5000 \times 50 \times 10^{-3} = j750 \, \Omega.$$

The ratio of the third harmonic to the first harmonic before using this circuit is

$$Attenuation_{before} = 100\frac{3}{5} = 60\%.$$

I'll use the voltage–divider relationship twice, once for a frequency of 5000 rad/s and once for a frequency of three times that. All I need are the magnitudes because they represent the sizes of the harmonics:

$$|A_{v1}| = \left| \frac{51 + Z_{L3}}{51 + Z_{L1} + 510 + Z_{C1}} \right| = 0.1388.$$

$$|A_{v3}| = \left| \frac{51 + Z_{L1}}{51 + Z_{L3} + 510 + Z_{C3}} \right| = 1.3254.$$

The attenuation of the third versus the first after passing through the filter, compared with the "before" values is

$$Attenuation_{after} = 100\frac{3|A_{v3}|}{5|A_{v1}|} = 573\%.$$

So the engineer's design has reduced the magnitude of the fundamental by a large factor (about 0.14) and increased the magnitude of the third harmonic by a modest amount (about 1.3).

If you are wondering how a passive circuit that has no amplifier can increase the magnitude of a voltage…. That's resonance at work!

3.8.3 Example XVIII

The circuit of Fig. 3.71 includes coupled coils. We are to find the output voltage $v_o(t)$ in the time domain after $t = 0$.

When we deal with coupled coils, mesh analysis generally yields equations that are easier to write than those using nodal analysis. I have chosen two mesh currents, i_1 on the left and i_2 on the right, both clockwise. The equation for the left mesh is

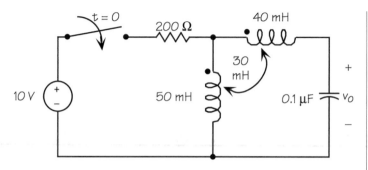

FIGURE 3.71: Example XVIII: Time domain response.

$$-10 + 200 i_1 + 50 \times 10^{-3} \frac{d\left(i_1 - i_2\right)}{dt} + 30 \times 10^{-3} \frac{di_2}{dt} = 0.$$

The first three terms are "standard" in the sense that they represent the three elements of the loop. The last term is the coupling term. Since i_2 is entering the dotted end of the 40-mH inductor, the induced voltage in the 50-mH inductor is positive at its dot.

The second mesh is a mess because the coupling appears twice. Current in the 40-mH induces voltage in that inductor and in the 50-mH inductor. Current in the 50-mH inductor also induces voltage in both inductors. The second mesh equation has five terms:

$$50 \times 10^{-3} \frac{d\left(i_2 - i_1\right)}{dt} + 40 \times 10^{-3} \frac{di_2}{dt} +$$

$$30 \times 10^{-3} \frac{d\left(i_1 - i_2\right)}{dt} - 30 \times 10^{-3} \frac{di_2}{dt} + v_o = 0.$$

The first two terms of the equation are the self terms for the inductors. The third term is the effect on the 40-mH inductor of current entering the dot of the 50-mH inductor ($i_1 - i_2$). The current enters the dot, making the other dot positive.

The fourth term represents coupling in the opposite direction. Since i_2 enters the dot, the voltage on the other coil is positive at its dot, which is opposite from our direction around the loop.

The last term is just v_o. I chose to do this instead of writing the integral relationship for the capacitor. Now I need to finish the equations by including the capacitor:

$$i_2 = 0.1 \times 10^{-6} \frac{dv_o}{dt}.$$

Finally, I need initial values of the currents through the inductors and the voltage across the capacitor. Since the circuit has presumably been dead for a long time before the switch is closed, these values are all zero.

$$i_1(0) - i_2(0) = 0,$$
$$i_2(0) = 0,$$
$$v_o(0) = 0.$$

The solution of these equations gives

$$v_o(t) = 3.956e^{-4051t} + e^{-702t}(0.626\sin 13782t - 3.956\cos 13782t)V \text{ for } t > 0.$$

3.8.4 Example XIX

Redo the analysis of the previous circuit, but this time in a sinusoidal setting. The circuit is shown in Fig. 3.72.

First, I need the impedances of the circuit elements:

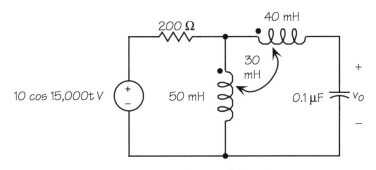

FIGURE 3.72:   Example XIX: Sinusoidal steady state.

$$\omega = 15,000,$$
$$Z_{L5} = j \times 15,000 \times 50 \times 10^{-3} = j750 \,\Omega,$$
$$Z_{L4} = j \times 15,000 \times 40 \times 10^{-3} = j600 \,\Omega,$$
$$Z_M = j \times 15,000 \times 30 \times 10^{-3} = j450 \,\Omega,$$
$$Z_C = 1/(j \times 15,000 \times 0.1 \times 10^{-6}) = -j666.7 \,\Omega.$$

With clockwise mesh currents I_1 and I_2 as before, my equations are

$$-10 + 200I_1 + j750(I_1 - I_2) + j450I_2 = 0,$$
$$j750(I_2 - I_1) + j600I_2 +$$
$$j450(I_1 - I_2) - j450I_2 - j666.7I_2 = 0,$$
$$V_o = -j666.7I_2.$$

Note the correspondence of terms between the previous example and this one. The first equation has three "self" terms and one coupling term. The second has three "self" terms and two coupling terms. Signs and directions are the same.

Solving these gives the phasor domain result of

$$V_o = 7.694 + j1.320 = 7.807\angle 9.7° \text{ V}$$

and a time-domain sinusoidal steady state result of

$$v_{o-sss} = 7.807\cos(15{,}000t + 9.7°)\text{V}.$$

3.8.5 Example XX

Find the Thévenin equivalent of the circuit shown in Fig. 3.73. Then find the correct matching load and the power delivered to that load.

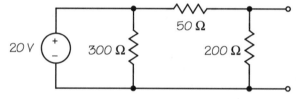

FIGURE 3.73: Example XX: Thévenin equivalent (DC).

I hope you noticed that the 300-Ω resistor is not part of the equivalent! The voltage across this resistor is 20 V and it contributes nothing to the output of the circuit—unless you consider heat a contribution.

The remaining circuit is a voltage divider, and the open-circuit voltage at the terminals is

$$v_{oc} = 20\frac{200}{200 + 50} = 16 \text{ V}.$$

The Thévenin equivalent resistance is the resistance "seen" from the terminals with the voltage source replaced by a short circuit. That resistance is the 50-Ω resistor in parallel with the 200-Ω resistor:

$$R_{Th} = \frac{1}{1/50 + 1/200} = 40 \,\Omega.$$

The matching load is therefore 40 Ω. Since the circuit is still a voltage divider, half the source voltage appears across the load. The power delivered to the load is

$$P_{match} = \frac{(v_{oc}/2)^2}{R_{Th}} = \frac{(16/2)^2}{40} = 1.6 \text{ W}.$$

3.8.6 Example XXI

Let's do a similar Thévenin example, but this time in the phasor domain. The circuit is shown in Fig. 3.74. We are to find the Thévenin equivalent, the matching load, and the power delivered to that load.

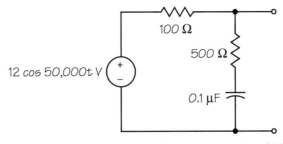

FIGURE 3.74: Example XXI: Thévenin equivalent (AC).

Convert the circuit elements to the phasor domain first:

$$\omega = 50,000,$$

$$Z_C = 1/\left(j \times 50,000 \times 0.1 \times 10^{-6}\right) = -j200\,\Omega.$$

Then use a voltage divider to find the open-circuit voltage in the phasor domain:

$$V_{oc} = 12\frac{500 + Z_C}{500 + Z_C + 100} = 10.2 - j0.6\text{ V}.$$

The Thévenin equivalent impedance is "seen" by "looking into" the circuit from the terminals with the voltage source replaced by a short circuit. This will be the 100-Ω resistor in parallel with the combination of the 500-Ω resistor and the capacitor:

$$Z_{Th} = \frac{1}{\dfrac{1}{500 + Z_C} + \dfrac{1}{100}} = 85 - j5\,\Omega.$$

The matching load is the complex conjugate of the Thévenin impedance:

$$Z_{match} = Z_{Th}^* = 85 + j5\,\Omega.$$

Finding the power delivered to the load is helped by realizing two things. First, the reactive parts of the impedances "eat up each other," canceling themselves out. This leaves only the two real parts. This is now a simple resistive voltage divider where the voltage splits between the two. The power delivered to the resistance of the load is the square of half the open-circuit voltage divided by the resistance. Here I use the magnitude of V_{oc} since the resistor is not interested in phase angles. The result is

$$P_{match} = \frac{\left(|V_{oc}|/2\right)^2}{\text{Re}\left[Z_{match}\right]} = \frac{\left(|10.2 - j0.6|/2\right)^2}{85} = 307.1\,\text{mW}.$$

3.9 CIRCUIT DESIGN EXAMPLES

One way to tie some of this together is to look at a couple of examples of circuits where we have a design goal. These aren't going to be complicated circuits, but they'll require us to think about what is needed and what we have in our toolbox.

3.9.1 Matching?

The circuit of Fig. 3.75 is an "academic" circuit, but it isn't much different from the kind of thing that you might find in, say, the model of a transistor amplifier. Our goal is to design the load R_L so that maximum power is delivered to that load.

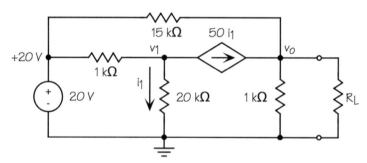

FIGURE 3.75: Circuit for matching.

Well, maximum power means Thévenin matching, provided we are allowed to change only the load. (If we can change the circuit itself, all bets are off.) So let's find the Thévenin equivalent as "seen" from R_L.

This circuit has a dependent source, so it will be best to find the open-circuit voltage and the short-circuit current. I'll write node equations, first for v_{oc}, the voltage at the terminals with R_L disconnected:

$$\frac{v_1 - 20}{1} + \frac{v_1}{20} + 50i_1 = 0,$$

$$-50i_1 + \frac{v_o}{1} + \frac{v_o - 20}{15} = 0,$$

$$i_1 = v_1/20,$$

$$v_{oc} = v_o.$$

The result is v_{oc} = 14.45 V.

Now I'll write node equations again, but this time with R_L replaced by a short circuit. Note that I can use the same set of equations by setting v_o = 0 and deleting the second equation:

$$\frac{v_1 - 20}{1} + \frac{v_1}{20} + 50i_1 = 0,$$

$$i_1 = v_1/20.$$

The result is $i_1 = 0.2817$ mA.

To find i_{sc}, which is the current flowing down through the short-circuit that replaced R_L, I will sum the current from the dependent current source and the current coming "over the top" through the 15-kΩ resistor:

$$i_{sc} = 50i_1 + 20/15 = 15.42 \text{ mA}.$$

Now find the Thévenin resistance:

$$R_{Th} = v_{oc}/i_{sc} = 0.938 \text{ k}\Omega = 938 \text{ }\Omega.$$

That yields the equivalent circuit that is shown in Fig. 3.76.

So the answer to our problem is to make $R_L = R_{Th} = 938$ Ω. The power delivered to this load (I know, we weren't asked for this, but…) is

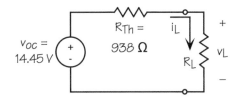

FIGURE 3.76:   Thévenin equivalent of fig. 3.69.

$$i_{L\max} = \frac{v_{oc}}{R_{Th} + R_{L\max}}, \quad P_{L\max} = i_{L\max}^2 R_{L\max} = 55.7 \text{ mW}.$$

Now we change our goal to require that we adjust the load resistor to make the voltage across the load 10 V. The Thévenin equivalent is still useful, because the question is again about the load. The circuit of Fig. 3.76 is a voltage divider, so "solving" the voltage divider for a 10-V output should do it:

$$10 = \frac{v_{oc}R_L}{R_{Th} + R_L}, \quad R_L = 2.105 \text{ k}\Omega.$$

This is a very different value from that required for maximum power transfer.

We have used a couple of our circuit-analysis tools in this design. The next example will stretch us a little more.

3.9.2 Interface Design

An "interface" is a circuit that we place between two other circuits to make them happy in some way. For example, we might use an interface to couple a low-impedance microphone into the high-impedance input of an amplifier.

In this design example, I want an interface that will allow me to connect a 2-kΩ load to the circuit we just dealt with and have 5 V delivered to that load. In other words, I want to get 5 V into a 2-kΩ load resistor using the circuit of Fig. 3.75 without modifying the original circuit.

As before, we are not changing the circuit to the left of the terminals, so I will use the same Thévenin equivalent to represent that circuit. Then the interface circuit, whatever it is, will connect between the terminals of the Thévenin equivalent and the 2-kΩ load resistor.

Well, what can I put into the interface? How about six resistors, a capacitor, two diodes, and a small elephant? No? Why not? Oh, so you think I could do a simpler one? Well, how about three resistors arranged in the shape of a T? No? Why not? Oh, I see, you think there's only one condition to meet so why not try one resistor first? Great idea!

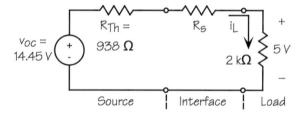

FIGURE 3.77: One-resistor series interface.

How would you like to arrange the resistor on the terminals? In series with the upper terminal? Fig. 3.77 is a possible circuit.

This is easy to solve by noting that the current i_L is established by the load requirements and then writing KVL around the whole loop:

$$i_L = 5/2, R_L = 2,$$
$$-v_{oc} + i_L(R_{Th} + R_s + R_L) = 0,$$
$$R_s = 2.844 \text{ k}\Omega.$$

So inserting that resistor in series with the load will do the job. But I'll bet at least somebody out there thought of putting one resistor in parallel as in Fig. 3.78.

I'll combine the 2-kΩ load and resistor R_p in parallel, then write the voltage–divider equation and solve for R_p:

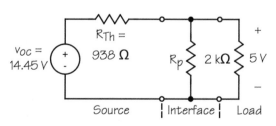

FIGURE 3.78: One-resistor parallel interface.

$$R_L = 2, R_{right} = \frac{R_L R_p}{R_L + R_p},$$

$$5 = v_{oc} \frac{R_{right}}{R_{right} + R_{Th}},$$

$$R_p = 0.66 \text{ k}\Omega = 660 \text{ }\Omega.$$

Which is the better solution, 2.844 kΩ in series or 660 Ω in parallel? That's hard to say without some criterion for measuring "better." I'll ignore the question and note that both solutions work. But there can be situations where a certain design won't work for some obvious reason, such as a negative resistor.

3.10   SUMMARY

Everything we have done in this chapter requires the circuit to be linear. Each of the rather special techniques gives us another way of looking at a circuit and finding out what is going on in it.

Some give more insight than others. Thévenin leads to matching, for example. Superposition gives us a way of seeing how one source among several influences a voltage or current. Reciprocity explains a few things that will be especially useful later in communications. Proportionality is a quick way to get simple results.

But out of all this stuff, what's important? Linearity and an understanding of what it means is at the top of my list. For number two I have to be wishy-washy; probably superposition and Thévenin's Theorem.

You will perhaps note that your tool box is getting heavier! We've got some better tools now, and we are going to use them in the rest of the course. In the next chapter we'll design interface circuits that go between a signal source and its load. And we'll need our new and improved tools.

CHAPTER 4

Designing an Interface: Some Practical Practice

What do we know at this point? Well, we have most of the basics of circuit analysis behind us. While there are still lots of things we could study, these basics form the foundation for all the rest. And if these are the foundation, we should be able to use them to create new circuits, not just analyze old ones.

"Create," of course, means "design." Each chapter has had a design problem at the end. In this chapter I am going to propose four design problems and then show how I arrived at a circuit.

Keep in mind, though, that the design that I do is not going to be the same as you or someone else might do. While there is usually only one "right answer" for analysis, in design there are many. In fact, one possible answer in design is "none of the above." We can develop problems that cannot be solved with the circuit theory that we know.

As you go through the designs in this chapter, think what you would do at each point, then see what I do. I'll bet you can find even better designs than I have, especially if you consider that we could have different definitions of the word "better."

4.1 AMPLIFIER LOAD

The transistor circuit of Fig. 4.1 is an amplifier that theoretically will amplify the incoming signal v_s by a factor of about 170. In practice the gain is probably more like 120, and then only for frequencies above a few hundred hertz.

We are to design an interface, shown on the right, that will make the 2-kΩ load "look" to the amplifier like a 600-Ω resistance. So the interface should somehow convert the load into, effectively, a 600-Ω resistor.

Hmmm, we don't know about transistors you say? OK, let's ask an expert how we can model this circuit so that we can consider the interface design: "Expert person, could you please help us make a model of this circuit?"

"Yes, indeed! But do you really need to know? After all, the design question relates entirely to the load and has nothing to do with the circuit that is driving the load. I'll be glad to help later, though, if you need it."

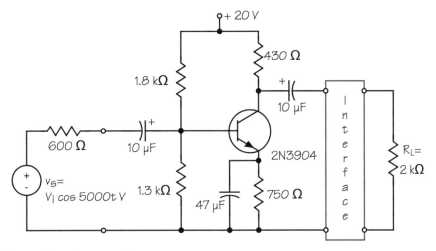

FIGURE 4.1: Transistor amplifer.

What should our interface look like? Here are some considerations:

- The interface should have as few parts as possible because we generally strive for simple designs that are cost effective and easy to understand and maintain.
- A resistor in series would merely add to the 2 kΩ already there. That takes us away from the 600 Ω that we want.
- A parallel resistor sounds like it might work because it would reduce the overall resistance.

Fig. 4.2 shows a possible interface.

All that remains is to realize that the parallel combination of R_p and the 2-kΩ load must yield 600 Ω. We can write and solve a simple equation:

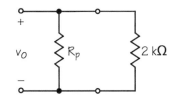

FIGURE 4.2: Possible interface.

$$\frac{2R_p}{2+R_p} = 0.6,$$

$$R_p = 857 \ \Omega.$$

Select a commercial 5% value[1], the nearest of which are 820 Ω and 910 Ω. I'll choose the 820-Ω resistor, which introduces an error of only about 3%. That's not unreasonable for much electronics work.

---

[1] 5% values begin with 10, 11, 12, 13, 15, 16, 18, 20, 22, 24, 27, 30, 33, 36, 39, 43, 47, 51, 56, 62, 68, 75, 82, and 91, with appropriate 10^n multipliers.

4.2 AMPLIFIER OUTPUT VOLTAGE

We've changed our minds a little. Not only do we want the amplifier's load in the previous example to be 600 Ω, but we want its output voltage to be 2 V (peak) when the input is 30 mV (peak).

It sure looks like we need some help with the amplifier now, doesn't it? Let's ask, "Expert, would you help us further now, please?"

"Of course! First, we need to know that we are interested in the circuit only for frequencies like the $\omega = 5000$ rad/s that is being applied in Fig. 4.1. At this frequency, the three capacitors must look essentially like short circuits. Why don't you check that first?"

OK, so we have our first job. Let's compute the capacitive reactance for each of these:

$$C_{in} = 10\mu F, \left|Z_{C_{in}}\right| = \frac{1}{\omega C_{in}} = 20\ \Omega,$$

$$C_E = 47\mu F, \left|Z_{C_E}\right| = \frac{1}{\omega C_E} = 4.3\ \Omega,$$

$$C_{out} = C_{in}.$$

Well, is the "short circuit" condition met? It's easiest to see for the capacitor that parallels the 750-Ω resistor. The capacitive reactance is 4.3 Ω and the resistor is 750 Ω. So the capacitor effectively short-circuits the resistor. The same is true for the other two capacitors.

"OK, Expert, we have verified that the capacitors act pretty much like short circuits. Now what? Can you make a model now?"

"Certainly! We'll make the *small-signal* model of this circuit. That's the model that ignores the DC sources and concerns itself only with the changing AC signal that is coming from the source v_s. I'll draw the model with all the right numbers for you in Fig. 4.3."

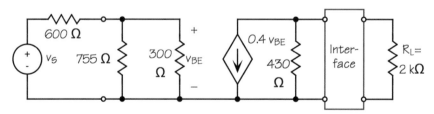

FIGURE 4.3: Small-signal model of Fig. 4.1.

"Gosh, is it that simple? Can we use this for our analysis and see what interface might work?"

"Oh, yes! The model is linear and it works pretty well if the signal v_s is kept small enough to avoid some transistor problems. So go ahead and analyze the model and design your interface. What you get will fit the requirements of the original circuit pretty well."

"Thanks, Expert!" Now that we are left alone, we will have to use the tools we know about. Is this a job for Thévenin? We care about the circuit only at the output end.

Yup, Thévenin equivalent. Combine the two parallel resistors on the left, then find v_{BE} using a voltage divider. That will allow us to analyze the right side of the circuit to find both the open-circuit voltage and the short-circuit current. From these we can get the Thévenin equivalent resistance.

The left side of the circuit reduces to

$$R_{parallel} = \frac{755 \times 300}{755 + 300} = 215 \ \Omega,$$

$$v_{BE} = \frac{215}{215 + 600} v_s = 0.264 v_s.$$

Find the output of the dependent source:

$$0.4 \times 0.264 v_s = 105.5 v_s \ \text{mA}.$$

Now that we know the current provided by the current source (note that it is downward), we can calculate the open-circuit voltage as

$$v_{oc} = -105.5 \times 0.430 = -45.4 v_s,$$

and the short-circuit current as

$$i_{sc} = -105.5 v_s \ \text{mA},$$

which means the Thévenin-equivalent resistance is

$$R_{Th} = v_{oc}/i_{sc} = \frac{-45.4 v_s}{-105.5 v_s} = 430 \ \Omega.$$

Now we can reduce the circuit to the very simple model shown in Fig.4.4.

Let's see, if the input voltage is 30 mV, the output with no load (the open-circuit voltage) will be

$$v_{oc} = -45.4 \times 0.03 = -1.36 \ \text{V}.$$

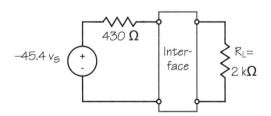

FIGURE 4.4: Simplified model.

That's hardly 2 V! What this says is that no resistive interface can do the job because we need some voltage gain.

Gain...? gain...? An op-amp, perhaps? The circuit shown in Fig. 4.5 will have gain and by choosing the input resistor to be 600 Ω, we can meet the condition that the load on the transistor amplifier is to "look like" 600 Ω. (Remember that the voltage between the − and + inputs is essentially 0, making the input look like a virtual ground.)

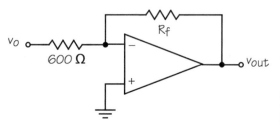

FIGURE 4.5: Possible interface.

Now connect this interface to our transistor circuit model and see what we need for the feedback resistor. Fig. 4.6 shows the connection.

Remember that the gain of this circuit is negative and is the ratio of the feedback resistance to the input resistance. Here, the input

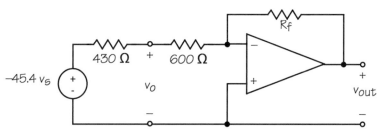

FIGURE 4.6: Model and interface.

resistance is the sum of the 430-Ω resistor of the model and the 600-Ω resistor on the minus input. The equations aren't difficult:

$$v_{out} = \frac{-R_f}{430 + 600}(-45.4v_s) = 2,$$

$$v_s = 0.030 \text{ V},$$

$$R_f = \frac{2 \times (430 + 600)}{45.4 \times 0.030} = 1512 \ \Omega.$$

I'll choose the nearest 5% resistor, which is 1.5 kΩ. But I also need to choose a commercial value for the input resistor. The nearest one to 600 Ω is 620 Ω. The final result is shown in Fig. 4.7.

Let's check the output v_{out} to make sure it meets the original specifications. After all, we've shifted the resistor values slightly:

$$v_{out} = \frac{-1500}{430 + 620}(-45.4v_s),$$

$$v_s = 0.030 \text{ V},$$

$$v_{out} = 1.95 \text{ V}.$$

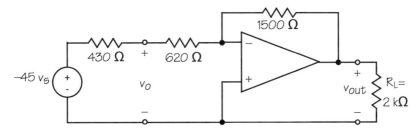

FIGURE 4.7: Complete interface.

That's close enough to 2 V! I'll take a 2.5% error in a circuit design almost any day.

4.3   LIGHT SENSOR

We have been given the job of designing a light meter with a digital output. One of our colleagues has developed a digital circuit that uses an analog-to-digital converter that has an analog input range of zero to ten volts. Our job involves interfacing the sensor and this A/D converter.

A cadmium sulfide photoresistor (Fig. 4.8) is available that has the resistance characteristic shown in Fig. 4.9.

The resistance characteristic of the CdS cell is a straight line when plotted on a log–log graph. The light intensity calibration is in lumens per square meter. You may have a feeling for foot-candles: 100 foot-candles is a brightly lit desk top. The conversion is

FIGURE 4.8: Cadmium sulfide photoresistor.

$$1 \text{ footcandle} = 10.764 \text{ lm/m}^2.$$

Most human activity is conducted between 100 and 1000 lm/m².

We are to provide a voltage using this CdS cell that will translate 2000 lm/m² into +10 V and 10 lm/m² into 0 V. We can read the CdS cell's resistance values from the graph in Fig. 4.9.

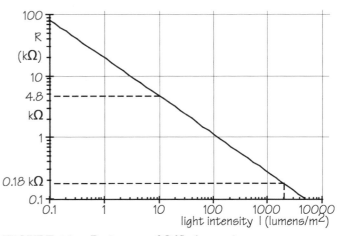

FIGURE 4.9: Resistance of CdS photoresistor.

$$R_{2000} = 0.18 \text{ k}\Omega,$$
$$R_{10} = 4.8 \text{ k}\Omega.$$

I think an op-amp is required. I'd like to be able to control the output voltage and the gain, and the op-amp seems suited for this. There are two places in a simple inverter that I can place the CdS cell:

- In place of the input resistor. Then as the light increases, the resistance decreases. Since this resistance is in the denominator of the op-amp's gain, the output would increase if I supplied the input of the op-amp with a constant voltage.
- In place of the feedback resistor. This would also change the gain, but the output would go in a direction opposite to the light intensity because as R_f goes up with decreasing light, gain goes down. I'll have to use the previous idea.

Figure 4.10 shows a possible circuit for doing this. I've used a fixed input of −10 V and placed the CdS cell in the input line.

I'll calculate the value of R_f so that the output at the maximum light intensity of +10 V:

$$v_{o\max} = 10 = \frac{-R_f}{0.18}(-10), R_f = 180 \ \Omega.$$

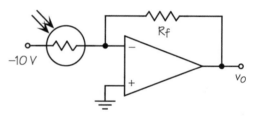

FIGURE 4.10: Possible circuit.

Check the output at the minimum light intensity:

$$v_{o\min} = \frac{-R_f}{4800}(-10) = 0.375 \text{ V}.$$

Oh, nuts, that's pretty far from being zero. I could ask the digital designer if that would be OK. No, better not, I'm new here and that wouldn't look good. But I could modify my circuit a little to include an offset. My op-amp summer is shown in Fig. 4.11.

Now I'll write two equations to define the resistances using the two light values, the corresponding cell resistances, and the required voltages:

$$v_{o\min} = 0 = \frac{-R_f}{R_1}(+10) + \frac{-R_f}{4800}(-10),$$

$$v_{o\max} = 10 = \frac{-R_f}{R_1}(+10) + \frac{-R_f}{180}(-10).$$

Solving these gives

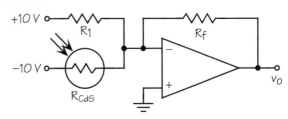

$$R_1 = 4800 \ \Omega, R_f = 187 \ \Omega.$$

FIGURE 4.11: Modified op-amp circuit.

That seems to do the job. Until we look at the values carefully. First, 187 Ω is a rather low value for the feedback resistor in an op-amp circuit using, say, an LM 741. At least 1 kΩ would be better. Second, the current in the CdS cell is rather high when the light is high. The cell resistance at 2000 lm/m² is 180 Ω, so the cell current is 10/180 = 56 mA. That seems high.

A bridge circuit often works in measurement. (I pulled that out of nowhere, didn't I!) We place the sensor in one arm of a bridge as shown in Fig. 4.12. I chose to place the CdS cell so that, when the light intensity increases and the resistance falls, the voltage on the minus terminal will also fall. This is inverted to give a rising output voltage.

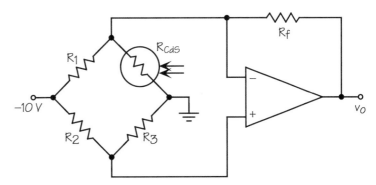

FIGURE 4.12: Bridge circuit for light sensor.

I choose the three resistors of the bridge so that they "fit" the resistance of the CdS cell in about the middle of its range. That middle is at about 2 kΩ, so I'll make all three resistors 2 kΩ each. (Why? Just a gut feeling, really, that seemed to make sense when I thought of it.)

The voltage on the plus terminal of the op-amp will be

$$v_+ = -10 \frac{2}{2+2} = -5 \ \text{V}.$$

Remember that the voltage between the input terminals of an op-amp is essentially zero. So $v_- = -5$ V also. Let's write and solve a KCL equations at the top of the bridge, using the maximum light intensity data:

$$v_- = v_+ = -5,$$

$$\frac{-10 - v_-}{2000} + \frac{v_{o\,max} - v_-}{R_f} - \frac{v_-}{R_{CdS}} = 0,$$

$$v_{o\,max} = 10,$$

$$R_{CdS} = 180.$$

The solution is $R_f = -594\ \Omega$! Oops, I must reverse something somewhere. I probably need a fixed input of +10 V rather than −10 V. The equations become

$$v_- = v_+ = 10\frac{2}{2+2} = 5\ \text{V},$$

$$\frac{+10 - v_-}{2000} + \frac{v_{o\,max} - v_-}{R_f} - \frac{v_-}{R_{CdS}} = 0,$$

$$v_{o\,max} = 10,$$

$$R_{CdS} = 180.$$

Now the solution is $R_f = 198\ \Omega$. That's better! But it is still pretty small. Moreover, calculating the output voltage for minimum light gives a voltage that's too high.

I'm going to try making the fixed input voltage +1 V rather than +10 V. That should make R_f much larger. Solving the same equations with 10 replaced by 1 and 5 replaced by 0.5, I get R_f = 3758 Ω. The nearest 1% resistor value is 3.74 kΩ, so I'll use that. The final circuit is shown in Fig. 4.13.

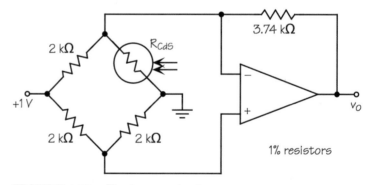

FIGURE 4.13: Final sensor circuit.

Oh, I forgot to check the low-light output. Let's see what it will be:

$$\frac{1 - 0.5}{2000} + \frac{v_{o\,min} - 0.5}{3740} - \frac{0.5}{4800} = 0.$$

The output $v_{o\,min}$ is −45 mV. If this is too large, we could tweak the values of the two 2-kΩ resistors in the lower part of the bridge to shift the voltage on the plus input just slightly to account for this. (A similar check of the output voltage at the maximum light values shows that it is about half of one percent low.)

I'll accept this design, knowing that I have some small calibration changes to make. In fact, I might want to add a couple of pots to make small adjustments. But that can be left until we try this circuit with the digital designer's A/D circuit.

Note that as I went through the design, I changed my approach several times. When you are designing something, this will often happen. You learn more about the problem so you try different approaches. While the choices you make are mostly discovered through your work, revelation sometimes takes a part.

4.4   HARMONIC SUPPRESSION

This last example involves changing the frequency characteristics of a signal. We often do this with signals. For example, a radio has several filters in it to remove frequencies that aren't part of the signal we wish to hear.

A certain signal is described mathematically as

$$v_s = 6\sin(2\pi \times 1000t) + 2\sin(2\pi \times 3000t) + 1.2\sin(2\pi \times 5000t) \text{ V}.$$

Recall that the terms that multiply t are in radians per second, which is 2π times the frequency in hertz. So the three terms are 1000 Hz, 3000 Hz, and 5000 Hz.

These terms are harmonically related, because the frequencies are integer multiples of the lowest frequency. We call the 1000-Hz term the *fundamental* (or first harmonic). The second term, at 3000 Hz, is three times the fundamental, so it is the *third harmonic*. Similarly, the last term is the *fifth harmonic*.

Fig. 4.14 is a plot of the signal over about a period and a half.

This signal v_s is produced by a source that has an internal impedance of 600 Ω. In other words, R_{Th} = 600 Ω.

Our goal is to design an interface to this source that will alter the signal by reducing the third harmonic to one-fifth the size of the fundamental (it is now one-third), while keeping the amplitude of the fundamental at 6 V.

Hmmm, what to do? Let's look at some possible ways to approach this problem:

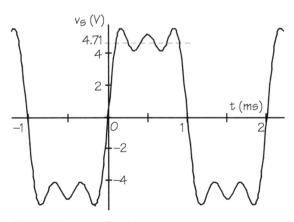

FIGURE 4.14:   Signal v_s.

- I can't make this out of just resistors, since their resistance is not a function of frequency (at least not in the audio-frequency range).
- An inductor in the top arm of a voltage divider might work, because the impedance of an inductor increases with frequency. The inductor would therefore add series impedance as the frequency goes up, and this impedance appears in the denominator of a voltage divider.
- A capacitor "across" the signal might work, because the impedance of a capacitor decreases with frequency. The capacitor would "short out" more and more of the signal as the frequency goes up.
- Capacitors are "better" elements in a number of ways. They are smaller and easier to make, they are fairly "pure" (an inductor has series resistance), and they can be made on silicon chips. I will try a capacitor first.
- My capacitor is likely to alter the fundamental as well, so I will probably have to add an amplifier to restore the fundamental to its original size.
- This added amplifier must not alter the circuit to which it is attached. In other words, I must arrange things so that the amplifier "loads" the capacitor as little as possible.
- It looks to me like an op-amp acting as a noninverter might do here. The input is to the plus terminal, so the input resistance is almost infinite.

All these thoughts have led me to the circuit shown in Fig. 4.15. I am using a capacitor (in combination with the source's internal resistance) to reduce the signal at v_1 as the frequency increases. I have an op-amp noninverter to correct the signal level.

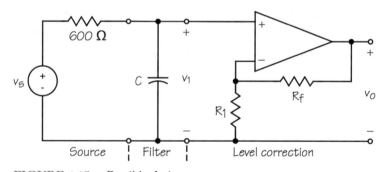

FIGURE 4.15: Possible design.

Now we must see what element values are needed.

Since the op-amp presents no load to the circuit at v_1 (recall that essentially 0 current flows into the + terminal), I can separate the circuit design into two parts: the source and filter, and the op-amp level correction.

My signal is the sum of three sinusoids. Only the fundamental and the third harmonic (the first two terms) are of interest, because that's what the specifications tell us to work with. Since these terms are two sinusoids at different frequencies, I will use superposition to reduce the circuit to things I know how to work with. Fig. 4.16 shows this.

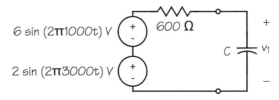

FIGURE 4.16: Sources superimposed.

How about writing a differential equation to analyze this circuit? If you are saying, "Ugh, no!" I'd agree with you. After all, we are being asked to modify the amplitude of the signal. Nothing is said about the transient response. So let's work in the sinusoidal steady state and save some effort.

But superposition can't superimpose signals of different frequencies in the sinusoidal steady state—unless you remember to change the impedance to account for the different frequencies. I will separate the sources in the time-domain circuit of Fig. 4.16, convert each circuit to its sinusoidal steady-state version at the proper frequency, and find the voltage v_1.

Consider first the circuit operating at the fundamental frequency of 1000 Hz, (2π 1000 rad/s). I'll draw the circuit with the source and the capacitor converted to the sinusoidal steady state. See Fig. 4.17. (The phase angle of $-90°$ came from starting with sine, not cosine.)

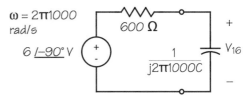

FIGURE 4.17: Circuit for fundamental.

This is a simple voltage divider, which yields the following result:

$$V_{16} = \frac{\dfrac{1}{j2\pi \times 1000C}}{\dfrac{1}{j2\pi \times 1000C} + 600} 6\angle -90°$$

$$= \frac{-j6}{1 + j1.2 \times 10^6 \pi C} \text{ V.}$$

Fig. 4.18 shows the same voltage divider, but with the 3000-Hz source applied. The result V_{12} is found in a similar way:

$$V_{12} = \frac{-j2}{1 + j3.6 \times 10^6 \pi C} \text{ V.}$$

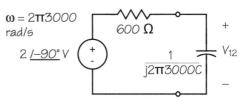

FIGURE 4.18: Circuit for third harmonic.

Now what's our goal? We need to find a value of capacitance that will make V_{12} 1/5 as large as V_{16}.

Ooops, no, we want the *magnitudes* to have the 1/5 relationship. So let's just write an equation that will do this:

$$|V_{12}| = \frac{2}{\sqrt{1 + \left(3.6 \times 10^6 \pi C\right)^2}}$$

$$= \frac{1}{5} \times |V_{16}| = \frac{(1/5)6}{\sqrt{1 + \left(1.2 \times 10^6 \pi C\right)^2}}.$$

Solving this (I used Maple, which required me to square both the terms before it would solve for C), I get

$$C = 0.142 \, \mu F.$$

Now we have a choice. A capacitor with a value of 0.142 μF is a rather rare beast. We could have a special one made, but our circuit would be expensive. Or we could choose the nearest commercial value. I am going to use the nearest commercial value and then check to see if my circuit performs as it should. That makes

$$C = 0.1 \, \mu F.$$

Before I check things, I'd like to get the output correct for the fundamental. It is supposed to be a 6 V peak. I'll see what $|V_{16}|$ is with C set to 0.1 μF. The result is

$$|V_{16}| = 5.61 \, V.$$

Hence the level correction must be

$$gain = 6/5.61 = 1.07 = 1 + \frac{R_f}{R_1}.$$

I tried various 5%-resistor values until I found

$$R_f = 1.1 \, k\Omega, \, R_1 = 16 \, k\Omega,$$
$$gain = 1.069.$$

Whew! That finishes the circuit, I hope.

But don't stop here! After all, I've chosen "imperfect" values (C, certainly). Does the circuit meet specifications? I need to check the final result.

Go all the way back to the calculations for V_{16} and V_{12}, then substitute the value of C and include the gain of the op-amp circuit. The result is

$$v_o = 6.00\sin(2\pi \times 1000t - 20.7°)$$
$$+1.42\sin(2\pi \times 3000t - 48.5°)$$
$$+0.60\sin(2\pi \times 5000t - 62.1°)\,\text{V}.$$

But the ratio is wrong, namely, 1.42/6.00 = 0.237! It's supposed to be 0.2. What have I done wrong? I know that $C = 0.142\ \mu\text{F}$ works. I adjusted to the nearest commercial value. But that made C smaller, and as the frequency rises, a smaller C has a larger impedance. I should have gone the other way:

$$C = 0.22\ \mu\text{F}.$$

Doing the same calculations yields

$$|V_{16}| = 4.62\ \text{V},$$
$$gain = 6/4.62 = 1.299 = 1 + \frac{R_f}{R_1},$$
$$R_f = 3\ \text{k}\Omega,\ R_1 = 10\ \text{k}\Omega,$$
$$gain = 1.30.$$

The final correct output is

$$v_o = 6.00\sin(2\pi \times 1000t - 39.7°)$$
$$+0.97\sin(2\pi \times 3000t - 68.1°)$$
$$+0.37\sin(2\pi \times 5000t - 76.4°)\,\text{V}.$$

Actually, the amplitude of the third harmonic (0.97) is less than one-fifth of the amplitude of the fundamental (6.00). That was the goal. Is this OK? Yes, for there's a margin, albeit rather large. But with commercial capacitors (often ±20%),

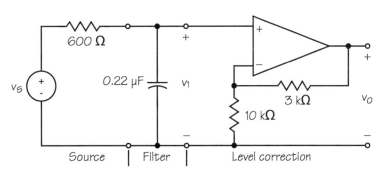

FIGURE 4.19: Final circuit.

that's OK. The final circuit is shown in Fig. 4.19 and the output is shown in Fig. 4.20.

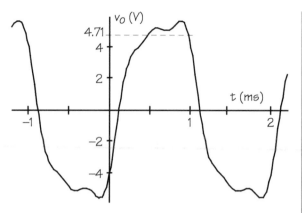

FIGURE 4.20: Final output.

Our circuit is a *filter* that modifies the incoming signal. You can see by comparing the input (Fig. 4.14) and the output that something certainly has changed! The signal is smoother, somewhat more like a sinusoid, because we have reduced the harmonics. But the signal is also delayed by our filter—note the late crossing of the time axis near $t = 0$.

I haven't been quite honest! First, our filter really consists not of just the capacitor but of the capacitor and the 600-Ω Thévenin impedance of the source. Second, there are much better ways to get such filters. In fact, there are rather simple designs that can suppress the harmonics by an even greater factor than one-fifth. Third, there are commercial capacitors with values between 0.1 and 0.22.

This is a good stopping point.

4.5 ANOTHER EXAMPLE

This example asks for a circuit with a strange (or at least different) twist. A DC voltage source of 1000 V is to produce 5000 V, at least for a moment. Circuits sort of like this are used in television sets, for example, to provide some of the high voltages required for the picture tube.

Here's the problem statement in more detail:

- We have a DC power supply that provides 1000 V. This supply can produce a maximum current of just 50 mA, however.
- We want some kind of a circuit that will deliver 5000 V to a load consisting of a 10-MΩ resistor and a 10-pF capacitor in parallel.
- The 5000-volt pulse doesn't have to last very long. All we need to do is get the capacitor charged to 5000 V and our job is done.

One way that comes to my mind is to "charge up" an inductor with a steady current from the power supply. I can limit that current by some series resistance. Then I will open the switch controlling the current. The capacitor–resistor load is directly across the inductor. So now I have the parallel combination of a "charged" inductor, a resistor, and a capacitor. I'll investigate the time-domain response of this circuit.

One problem in designing the circuit is how to build the inductor. The inductor is made of wire wrapped into a coil. That wire has resistance: the longer the wire, the higher the resistance, and the thicker the wire, the lower the resistance.

To simplify this example, I am going to assume that the wire I use and the form of the coil are such that the coil resistance is 500L Ω, where L

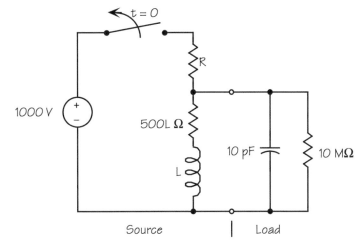

FIGURE 4.21: Circuit design.

is the inductance of the coil in henries. (If we wanted to change the inductor's design later, we could include R_L as a parameter.)

A possible circuit for doing this is shown in Fig. 4.21. The resistance R will be chosen later to limit the current to 50 mA with the switch closed.

I'll start my design by analyzing the time-domain response of this circuit after $t = 0$, assuming that the initial inductor current is 50 mA. The inductance L will be a parameter. Then I will find a value of L that will give a voltage peak of 5000 V across the capacitor.

I have chosen to write Kirchhoff's current law at the top node (after the switch opens). But I'll avoid the integral for the inductor by writing just i_L and then using a second equation to give v_o the voltage across the load in terms of i_L.

The initial capacitor voltage is determined by the steady 50-mA current flowing through the inductor's resistance 500L.

$$i_L + 10 \times 10^{-12} \frac{dv_o}{dt} + \frac{v_o}{10 \times 10^6} = 0,$$

$$v_o = L\frac{di_L}{dt} + (500L)i_L,$$

$$v_o(0) = (500L)i_L(0),$$

$$i_L(0) = 50 \times 10^{-3}.$$

Now comes the fun! Maple can produce a solution for $v_o(t)$ for $t > 0$, but it is a big mess. However, even though the result is a mess, Maple often happily plots the result.

I plotted the result for different values of L. The waveform was always a damped sinusoid. With L = 120 mH, the largest peak of the sinusoid is −5400 V. That satisfies the requirement. It also establishes the value of the inductor's resistance.

$$L = 120 \, \text{mH},$$
$$R_L = 500L = 60 \, \Omega,$$

From this information, I can get the value of the current-limiting resistor:

$$-1000 + (R + R_L)i_L(0) = 0,$$
$$R = 19,940 \, \Omega.$$

I won't plot the result here, but it's a damped sine wave that starts negative, peaks at −5400 V, and has a period of 6.9 μs. The exponential decay appears to have a time constant of about 180 μs.

4.6   SUMMARY

We've gone through several problems that can be classified as interface designs. Our goal in each is to accept a signal that has certain specifications and do something with it to get the output signal:

- In Section 4.1 the condition was to make the load appear to be a certain impedance.
- In Section 4.2 we were to provide a certain level of output signal while meeting some impedance restrictions.
- In Section 4.3 we connected a sensor that provided to an analog-to-digital converter a signal related to light intensity.
- In Section 4.4 we filtered a signal to reduce certain a component of it in the frequency domain.
- In Section 4.5 we created a high-voltage pulse into a given load.

So what's important? It certainly isn't the results of these designs. No, it's more how to approach a design problem. It's the idea that, when you are confronted with lots of unknowns, you need to look first at the simpler things you know.

How do you know what you know? There's no simple answer. You look at the problem. Then you look back at the circuits you've worked on. Which ones seem to have the correct general characteristics? Try one, using analysis techniques that give you insight quickly. And when you've got one that looks pretty good, go into a more detailed analysis to prove (or disprove) your choice. Trial and error? Yup, but with insight built on both experience and education.

What's next? This is the end of *Pragmatic Circuits I—DC and Time Domain. Pragmatic Circuits II—Frequency Domain* will shift focus to working directly with frequency as the independent variable rather than time. We are going to be more concerned with the frequencies of signals and what happens to them, rather than the shape of the signal itself as a function of time. This second part will also include AC power. All of this puts us in the sinusoidal steady state, where the sine wave is now the king.

Biography

Bill Eccles has been Professor of Electrical and Computer Engineering at Rose-Hulman Institute of Technology since 1990 (except for one year at Oklahoma State). He retired in 1990 as Distinguished Professor Emeritus after 25 years at the University of South Carolina. He founded the Department of Computer Science at that university, and served at one time or another as head of four different departments, Computer Science, Mathematics and Computer Science, and Electrical and Computer Engineering, all at South Carolina, and Electrical and Computer Engineering at Rose-Hulman. Most of his teaching has been in circuits and in microprocessor systems. He has published Microprocessor Systems: A 16-Bit Approach (Addison-Wersley, 1985) and numerous monographs on circuits, systems, microprocessor programming, and digital logic design. Bill learned circuit theory at M.I.T. under Ernest Guillemin, one of the pioneers in modern circuit theory, and William Hayt at Purdue University. Bill and his wife Trish have two children and three grandchildren. Bill is also a conductor (appropriate for an electrical engineer) on the Whitewater Valley Railroad, a tourist line in Connersville, Indiana. He is a Registered Professional Engineer and an amateur radio operator.

Printed in the United States
by Baker & Taylor Publisher Services